...icy for national forests in the eastern United States

the lands nobody wanted

A Conservation Foundation Report

by William E. Shands and Robert G. Healy

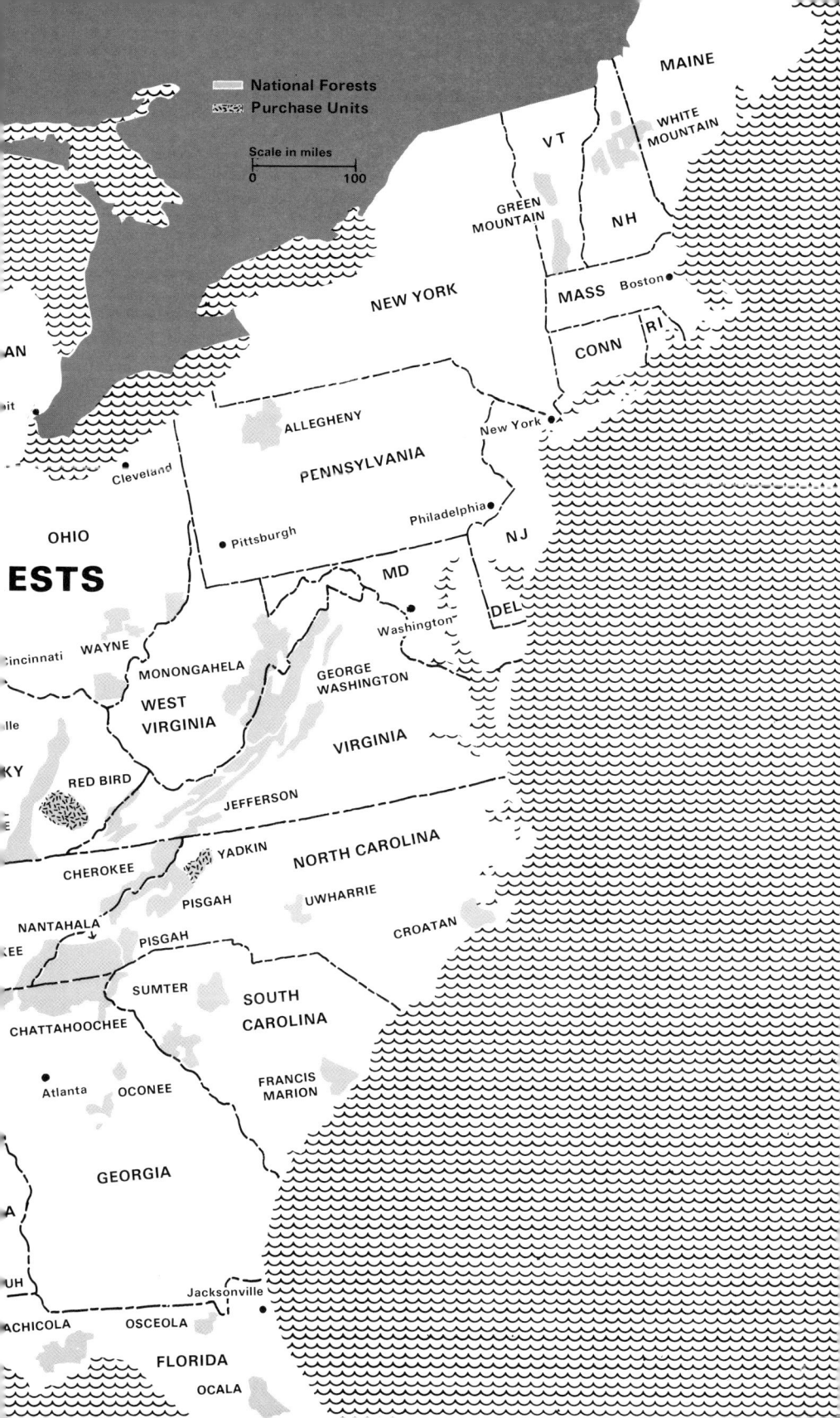

the lands nobody wanted

Policy for national forests in the eastern United States

the lands nobody wanted

A Conservation Foundation Report

by William E. Shands and Robert G. Healy

THE CONSERVATION FOUNDATION

The Conservation Foundation BOARD OF TRUSTEES

Ernest Brooks, Jr., Chairman   William H. Whyte, Jr., Vice-Chairman

John A. Bross	Philip G. Hammer	James W. Rouse
Louise B. Cullman	Walter E. Hoadley	William D. Ruckelshaus
Dorothy H. Donovan	William T. Lake	Anne P. Sidamon-Eristoff
Maitland Edey	Richard D. Lamm	James Hopkins Smith
Charles H.W. Foster	Lord Llewelyn-Davies	Barbara Ward (Lady Jacksc
David M. Gates	Cruz Matos	Pete Wilson
Charles Grace	David Hunter McAlpin	George M. Woodwell
D. Robert Graham	Tom McCall	
Nixon Griffis	Richard B. Ogilvie	William K. Reilly, Presiden

© 1977 by The Conservation Foundation
1717 Massachusetts Avenue, N.W., Washington, D.C. 20036
Second Printing, March 1979

Library of Congress Catalog Card Number: 77-72771
International Standard Book Number: 0-89164-042-8 (hardbound)
0-89164-043-6 (paperbound)

All rights reserved. No part of this book may be reproduced in any form without the permission of The Conservation Foundation.

The quotation on pages 100-101, from Ann and Myron Sutton's *Wilderness Areas of North America*, copyright © 1974 by Funk and Wagnalls Publishing Co., Inc is reprinted with permission of Thomas Y. Crowell Company, Inc.

All photographs are by Janet Mendelsohn except those on pages xxiv, 23, 54, 98, 114, 118, 139, 144, 175 (top), 183, 241, 258, 268, which are courtesy of the U.S. Fore Service.

design/production: Leckie/Lehmann

Printed in the United States of America.

THE CONSERVATION FOUNDATION AND PUBLIC LANDS

The Conservation Foundation is a nonprofit research and communication organization dedicated to encouraging human conduct to sustain and enrich life on earth. Since its founding in 1948, it has attempted to provide intellectual leadership in the cause of wise management of the earth's resources. It is now focusing increasing attention on one of the critical issues of the day—how to use wisely that most basic resource, the land itself.

From its beginning, the Foundation has encouraged the protection and enhancement of the most bountiful lands owned by the American people —the national parks and national forests. A series of Foundation studies and publications examined the complex relationship between man and nature in parks and forests, and the impacts on these lands from the relentless pressures of an increasing population, higher material standards of living, and increased leisure time. In 1972, at the request of the National Parks Centennial Commission, The Conservation Foundation conducted a broad examination of the past, present, and future of the National Park System, resulting in the publication of *National Parks for the Future*, which recommended priorities and direction for future park management. More recently, it initiated a program of citizen participation in planning the new Cumberland Island National Seashore in Georgia and, at the request of the National Park Service, examined management policies and land-use conflicts at Fire Island National Seashore in New York. In these and other projects, the Foundation has tried to render an impartial and independent assessment of management direction, recognizing that diverse interests make legitimate demands on public lands. In developing recommendations for management policies, the Foundation seeks to reconcile and accommodate these demands to the extent consistent with the capacity of the resource base.

THE AUTHORS

WILLIAM E. SHANDS, a senior associate with The Conservation Foundation, is a policy analyst and specialist in land use and public lands policy. He is a former journalist and held positions as an administrative and legislative aide in the U.S. Senate and House of Representatives. Since joining the Foundation in 1972, he has pursued a continuing interest in public lands policy and management and in the resolution of conflicts between the uses of public land—parks, forests, and wildlife refuges—and adjacent private land.

ROBERT G. HEALY, an economist and land-use specialist, is a senior associate with The Conservation Foundation. Before joining the Foundation staff in 1975, he was a member of the research staffs of the Urban Institute and Resources for the Future. He also has lectured on city planning at Harvard University. His book, *Land Use and the States*, a study of state involvement in land-use control, was published by Johns Hopkins University Press for Resources for the Future in 1976. He also directed the Foundation's study of the California Coastal Zone Conservation Commissions.

contents

Foreword xi
Acknowledgments xix
I Backyard to Megalopolis 1
II Present Uses and Future Demands 25
III Objectives, Opportunities, and Incentives 79
IV Forest Legislation and the Forest Service 119
V Opportunities for Cooperation and Coordination 157
VI Ownership Patterns and Acquisition 199
VII The Forests of the Future 235
Appendix A: Statistical Profile of the Eastern National Forests 265
Appendix B: Participants in Regional Workshops 269
Index 277

TABLES
1 A Comparison of Eastern and Western National Forests 6
2 Current and Projected U.S. Consumption of Timber 38
3 Outdoor Recreation in Eastern National Forests (1974) 42
4 Projected National Demands for Outdoor Recreation 50
5 Relative Preference of Various Socioeconomic Groups for Outdoor Recreational Activities 71

MAPS
Eastern National Forests endsheets
National Forest System xxii-xxiii
Planning Regions for Eastern National Forests 133
Marietta Unit of the Wayne National Forest 198
White Mountain National Forest 237
Ocala National Forest 245
Allegheny National Forest 255

foreword

Forests cover a third of the nation. If we exclude Alaska and Hawaii, 60 percent of our total forested land lies in the densely populated region east of the Mississippi. In fact, some of our most urbanized states—including New York and New Jersey—are more than half-forested.

Most eastern forest land is privately owned. However, there are 50 national forests in the East which constitute the largest public land system east of the Rocky Mountains. While they amount to only six percent of all the forested land in the East, they represent scenic, timber, wildlife, mineral, and recreational resources of enormous importance.

It was not always so. The vast and magnificent wilderness that greeted early American settlers was used to provide timber for their wheels and walls, bridges and barns. As the East industrialized and urbanized, its forests declined in size and quality. Wildfire consumed the forest debris left behind; rain swept away the topsoil. By the end of the 1800's, the primeval forest in the eastern United States had virtually disappeared. The brush and tree cover that eventually grew back was a poor substitute for the climax forest it replaced.

As forest devastation pushed westward, concern increased that the once-great American forest would disappear entirely. In the West, in an effort to control wholesale private exploitation of remaining forests, the Federal Government reserved vast areas of public domain land. The purpose, as set out in the Organic Act of 1897, was to protect water flow from the forests and to provide "a continuous supply of timber for the use and necessities of citizens of the United States."

But at the turn of the century, national forests, then called forest reserves, existed only in the West. Little public domain land remained in the more heavily settled East; if forests were to be established there, land would have to be purchased by the Federal Government.

In 1911, with passage of the Weeks Law, which authorized the Federal Government to buy land, Congress determined that the East should have federal forests, particularly in the southern Appalachian Mountains and the White Mountains of New Hampshire. Later, during the Great Depression, the prospect of federal forest protection and rehabilitation—and federal purchase dollars—became attractive in other regions as well. In the 66 years since passage of the Weeks Law, 50 national forests totaling almost 24,000,000 acres of federally owned land have been established in 23 eastern states. It is these forests with which this report is principally concerned.

DISTINCTIVENESS OF THE EASTERN NATIONAL FORESTS

The eastern national forests differ markedly from the larger forests west of the 100th meridian. The 24,000,000 acres in the East amount to only 13 percent of the total National Forest System. On a map of the continental United States (see page xxii - xxiii), the western forests appear as bold blocks of land overlaying high mountain ranges, while the eastern forests seem to be flecks scattered across their half of the nation.

Because they were purchased, not carved from large blocks of public land, the eastern forests are characterized by a fragmented ownership pattern, creating a patchwork of public and private lands. In fact, only about 51 percent of the land within the designated boundaries of the national forests of the East is federally owned. In one forest, federal ownership is only 20 percent. Nevertheless, these forests offer significant potential for satisfying local and regional environmental, economic, and recreational needs. Some 174,000,000 people live and work within a day's drive of an eastern national forest—an indication of the intensity of pressure upon this relatively small land system.

Yet attention usually focuses on the huge western forests, often rich in mature timber and abounding in roadless wilderness. To many policy makers, Forest Service officials, and conservationists, "the West is where the action is." Policies developed for the vast, consolidated western lands are applied nationwide. Too often, these policies deal inadequately with the special circumstances of the East.

Concerned that the eastern national forests were not receiving the special attention they need, The Conservation Foundation late in 1974 initiated a public policy evaluation of these lands. Principal financial support was provided by The Andrew W. Mellon Foundation. Additional help came from a number of other sources, including The Nature Conservancy.

As a major component of this study, we sponsored four two-day regional workshops—at Waterville Valley in the White Mountain National Forest (for New England); at Nemacolin Inn in Pennsylvania (for Pennsylvania, Ohio, and West Virginia); at Airlie House near Washington, D.C. (for Washington-based representatives of forest interests); and in Atlanta, Georgia (for the Southeast). Almost two hundred people participated in these discussions. They included a broad variety of forest interests—representatives of timber, energy, recreation, and land-development industries; professional foresters; members of environmental and public-interest organizations; farmers; hunters; state and local officials; and small landowners. The workshops, designed to help us gain the perspectives of diverse forest interests, proved a valuable source of ideas. Many of these are reflected in the pages that follow.

The purpose of this report is to call attention to the distinctive qualities of the eastern national forests and the various demands placed upon them, to stimulate public discussion about forest uses and policy, and to encourage long-range management policies and initiatives sensitive to these lands. It is intended for many audiences: the Congress, which has the authority to implement many of the recommendations; the Forest Service; state and local officials; those who use the forests directly, either for work or recreation; and particularly the millions of Americans, most of whom live in urban areas, who do not realize how the national forests of the East affect them.

PRINCIPLES FOR MANAGING THE NATIONAL FORESTS OF THE EAST

National forests are not parks. Their historic role as "working forests" is their special excitement. But in the effort to increase their productivity—for recreation as well as timber—we must not permit their degeneration into low-grade, disturbed environments lacking in distinction. That would be tragic. The work of the Forest Service in rehabilitating the eastern national forests—in large measure land that only recently nobody wanted—is one of the great conservation achievements of American history. It is also one of the great success stories of federal concern and action, an example worth emphasizing in a time of widespread cynicism about the wisdom and capacity of the Federal Government. To build upon this achievement, **we recommend that future management of the eastern national forests for the long-term benefit of society give priority to two basic principles:**

- **First, to providing public benefits that cannot be supplied by private land, either because resources are unavailable or because an economic incentive is absent.** As a New England conservationist put it, "If we're going to manage the eastern national forests just like all the other land, then we might as well sell them off."
- **Second, to restoring the forests as natural environments, distinct from the man-made environments otherwise dominant in the East. The forests and their products should be used only to the extent that this continuing process of restoration is not interrupted.**

These simple principles have profound implications. Adherence to them would clear away much of the fuzziness from forest policy debate. They provide a solid foundation for determining priorities and for making decisions on forest programs. Explicitly, they counter the unachievable objective of "something for everybody, everywhere, on the public's forest lands."

Applying these principles to timber, **we recommend that the eastern national forests complement the production of the far more extensive private forests in the East, which should continue to provide the bulk of the nation's hardwood timber and should also provide a significantly greater portion of the softwood than at present.** Economic circumstances are leading private timber growers to convert to softwoods and to emphasize pulpwood production. So that the public lands will fulfill needs not met by private lands, **the eastern national forests should stress quality, specializing in superior hardwood and softwood sawtimber, with long growing cycles.** This will ensure to future generations the hardwood material they will require for furniture and veneer. We favor longer growing periods on much of the eastern national forest land for several reasons—they're necessary for the hardwoods, good for recreation and the natural environment, and government alone seems able to afford them.

For recreation, we recommend that the national forests specialize in those types of recreation—usually dispersed, low-intensity activities—which require large areas of relatively natural terrain. High-intensity uses—those most likely to be commercially profitable—should be diverted to private lands. As in the case of timber, the public and private lands should complement one another, with public lands providing the resource—the wild rivers, rugged mountains, and wildlife—and private lands offering developed facilities at a fair price to the user, with a fair return to the entrepreneur.

We recommend that modest fees be levied on recreationists in the national forests, both to obtain revenue from those who use and enjoy the forests directly and to demonstrate to decision makers the intensity of the demand for high-quality recreation. And we suggest the creation of certain financial incentives to encourage local landowners to manage their lands profitably for timber and recreation, thus reducing demands on the national forests.

To promote the restoration of these forests as natural environments, we also suggest establishing low management intensity areas—what might be considered "temporary wildernesses." These would be relatively remote areas where timber would be managed on very long rotations and cut at the end of the growing cycle. For quality hardwoods, this will frequently mean more than a century during which no timber cutting is necessary. These areas, established in addition to, and not as a substitute for, true, designated wilderness, would satisfy many of the needs of people who desire a scenic experience, but who would be satisfied with something less than true wilderness.

The basic principles we suggest for eastern national forest management require that new attention be directed to the environmental disruptions that mining is causing on the forest surface. The Forest Service owns the mineral rights under only about two-thirds of its land, with the proportion of private ownership highest in the areas containing known mineral deposits. In cases where the ownership of minerals remains in private hands, the Forest Service has limited or no control over the extraction of minerals from beneath its forests.

Mining, the most land-intensive of all the forest uses, could affect as much as 500,000 acres of national forest land in the East. Already, oil, gas, coal, and other minerals are being extracted from beneath these forests. Recent price increases in fossil fuels have prompted energy companies to take another look at mineral deposits once considered uneconomic to mine, in such places as the old oil fields beneath the Allegheny National Forest in Pennsylvania. Surface mining for coal currently threatens some of the most wild and beautiful areas of the Monongahela National Forest in West Virginia.

We propose a systematic action program to protect the forest surface from mining damage, primarily through reliance on state and federal regulatory authority, with acquisition as a last resort. We also urge that the Forest Service aggressively exert its legal rights as a surface owner or adjoining property owner. In addition, we suggest that the Forest Service assume the responsibility for reclaiming abandoned strip-mined land, since much of it would most appropriately be restored to forest use.

LEGISLATIVE NEEDS

The special role of the eastern national forests should be recognized by statute, not by administrative action alone. We recommend the enactment of legislation which explicitly recognizes the distinctive characteristics and values of these eastern forests and their role in the total land system. The statute we suggest would establish, as the management mandate for these forests, the provision of renewable resources of superior quality and variety such as the private sector will not or cannot provide.

While this report focuses on the eastern forests, we have suggested some legislative and institutional changes that have applicability and benefit to the entire Forest Service system. One is the recommendation that a new use category—the protection and enhancement of natural amenity and environmental values—be established, to make explicit the parity of natural or amenity uses and values with timber, recreation, and the other uses listed in the Multiple-Use Sustained-Yield Act. We also recommend that the Forest Service, by statute, be given a charge that stresses its continuing responsibility for the healthy functioning of forest resources under its management. And we suggest a change in the wording of the Organic Act to strengthen the present protection phrase to make forest restoration, protection, and improvement a fundamental forest purpose.

INTERDEPENDENCY OF PUBLIC AND PRIVATE LANDS

Fragmented ownership patterns increase the interdependency of the eastern national forests and adjacent private lands. Incompatible uses of adjacent land call into question the very manageability of some eastern forests and profoundly threaten environmental quality in many more. Thus, the second-home developments that girdle Massanutten Mountain in the George Washington National Forest have already eroded the natural character of the mountain; the same situation is evident in the Highlands area of the Nantahala National Forest in North Carolina. In Arkansas, a developer wants to build hunting cabins on a 160-acre tract in the middle of a national forest game management area. In the Nicolet National Forest in Wisconsin, lakefront landowners want a more direct road across the national forest land to their holdings. At the same time, they oppose Forest Service plans to improve public access to the lake, apparently unaware of the irony of their position. One could cite dozens of conflicts between the uses of the public's land and the uses, desires, and expectations of private owners.

One way of defending the public's land is by purchasing private parcels liable to be developed in incompatible ways. In certain situations—that 160-acre inholding in a forest game area is a good example—acquisition can be an effective and appropriate response. **We recommend that funds for general forest acquisition be increased nearly seven-fold above present levels, to $10,000,000 annually, to acquire strategic inholdings and other lands of high resource value—particularly where the Federal Government is the only agency able and willing to acquire and manage land on the scale required.**

Even with increased acquisition, large areas of private land will remain within forest boundaries. There will never be enough money to "block up" all the public's land into large, consolidated holdings. Nor is solid federal ownership necessarily desirable in all cases. Scattered ownerships in some forests ensure a diverse landscape, allow continued farming on valley floors, and accomplish other desirable land-use objectives.

Since the pattern of patchwork ownership will continue, the publicly owned lands within the eastern national forests simply cannot be managed in isolation. Defensive acquisition and policies and programs that do not recognize interdependencies will ultimately fail to realize the full benefits from the public's lands. The national forests, if not managed with sensitivity to their surroundings, can impose burdens on their neighbors. Nor can those vested with authority over the private land—either as public officials or landowners—shrug off consideration of their neighboring national forests. Impacts flow both ways, so it is to everyone's benefit to have land uses coordinated and complementary. **More cooperative relationships must be developed between the public land managers and state and local officials who regulate private land, and the landowners themselves.**

Recent federal legislation providing for an increased, and assured, minimum per-acre payment to counties for national forest land will eliminate a major friction point between the Federal Government and local governments. **To help the Forest Service take the lead in developing cooperative programs and activities, we recommend a number of actions, including improvements in the national forest planning process; the addition to the Forest Service staff of personnel knowledgeable in planning, economics, and other aspects of community development; and stronger linkages between national forest staffs and Forest Service personnel of the State and Private Forestry branch to increase and improve technical assistance to governments and landowners in the forest environs. We suggest that state and local governments exercise their authority to control land use around the forest, perhaps through new institutional arrangements that draw on federal, state, and local authorities and programs.**

Decisions affecting the future of these forests should consider their potential for meeting local, regional, and national needs. The national forests and their associated resources are components of a much larger land system—in this case, the United States east of the 100th meridian. All of the resources, whether timber or wildlife habitat, wilderness or watershed, whether owned by the public or by private individuals, are integral parts of the larger system and each must play its own particular and appropriate role. The identification of that special role is the challenge of this work— and of national land and resource policy. In essence, we agree with a participant in our New England forest workshop: "We can write our own future for our national forests; we don't have to simply drift and see what happens."

The national forests of the eastern United States are a Cinderella story. Plucked from dire straits, treated with care, they have blossomed beyond the imaginings of those who created them. But the coach is waiting, the clock is ticking, and it's time for action.

WILLIAM K. REILLY, *President*
The Conservation Foundation

acknowledgments

In preparing this book, we were aided immeasurably by the ideas and advice of scores of individuals. Many in the Forest Service welcomed this outside review of Forest Service policy, programs, and management. Essential to us in launching this project in 1974 were the strong support and encouragement of Chief John R. McGuire; Jay H. Cravens, former regional forester, Eastern Region; and R. Max Peterson, former regional forester, Southern Region (now deputy chief, programs and legislation). Forest Service personnel in Washington and the regional offices, as well as forest supervisors and district rangers, patiently answered our questions and supplied data without hesitation. Raymond M. Housley, Jr., associate deputy chief of the Forest Service, was assigned liaison duties at the project's inception, and his interest and responsiveness never flagged. Invaluable insights also were provided by John A. Sandor, former deputy regional forester, Eastern Region (now regional forester, Alaska); Kenneth C. Scholz, former assistant director, lands, now with the Forest Service's Surface Environment and Mining Research Project in Billings, Montana; and Michael J. Penfold, Supervisor of the Thomas Jefferson National Forest. Roy C. Gandy and Carl N. Wilson served as official Forest Service liaisons for the southern and eastern regions, respectively, and provided data and technical review of drafts. During our visits to the forests, forest supervisors were most cooperative and informative, and district rangers were especially helpful in increasing our understanding of on-the-ground management issues.

Three regional environmental organizations cooperated with The Conservation Foundation throughout the project. Each convened a regional workshop to consider forest management issues and options (a list of workshop participants appears on pages 269-275), prepared regional discussion papers and summaries of the workshop proceedings, reviewed drafts of the manuscript, and assisted particularly in the preparation of Chapter VII. We extend special thanks to Joshua C. Whetzel, Jr., president of the Western Pennsylvania Conservancy, and Arthur A. Davis, director of the Conservancy's Pennsylvania Land Policy Project; Perry R. Hagenstein, executive director, New England Natural Resources Center; and William M. Partington, executive director, Florida Environmental Information Center.

Robert C. Dennis, while a senior associate with The Conservation Foundation, developed the idea for a study of the eastern national forests and worked on its early conceptualization. We were fortunate to have, for too-brief periods of time, the services of two very capable research assistants, Cary B. Hinton, whose work was partially supported by an intern grant from the Jessie Smith Noyes Foundation, and Margaret Coon. Mr. Hinton analyzed and prepared papers on the cooperative activities of the Forest Service and Ms. Coon collected data on the complex minerals situation and conducted a case study of land use in and around the forests in the Southern Appalachians. Katherine Forbes also served as a summer intern. A useful case study of development around the Green Mountain National Forest was prepared by Lloyd Irland and Thomas Siccama of the Yale School of Forestry and Environmental Studies. William J. Duddleson of the Foundation staff contributed a study of Sawtooth National Recreation Area.

The photographs taken by Janet Mendelsohn and the interviews she conducted with forest users in the east add a dynamic and personal dimension to this study of forest policy. Continuing counsel throughout the preparation of this report was provided by J. Clarence Davies, III, and John H. Noble, vice presidents of The Conservation Foundation. We also thank other members of the Foundation staff who provided us with information in their respective areas of specialization. And we gratefully acknowledge the capable editing of the manuscript by Janet M. Fesler of the Foundation.

A number of people reviewed early drafts of the report and offered helpful comments. Among them were: Marion Clawson and William Hyde, Resources for the Future; Edward Cliff, former Chief of The Forest Service; Charles H. W. Foster and David M. Smith, Yale University School of Forestry and Environmental Studies; Maitland Sharpe, Izaak Walton League of America; Richard Pardo, American Forestry Association; Dennis LeMaster, Society of American Foresters; John R. Castles, formerly of the U.S. Forest Service; and Barry R. Flamm and Larry W. Hill, Council on Environmental Quality.

In addition, we received useful comments on individual chapters from Boyd Gibbons, National Geographic Society; Hamilton Pyles, Natural Resources Council of America; John S. Gottschalk, International Association of Game, Fish, and Conservation Commissioners; John Rosenberg, of the Foundation staff, and many others. We always welcomed their suggestions and, more often than not, concurred with them. The book is much the stronger for their assistance, but they, of course, bear no responsibility for the final product. The findings and recommendations are those of The Conservation Foundation.

William E. Shands
Robert G. Healy

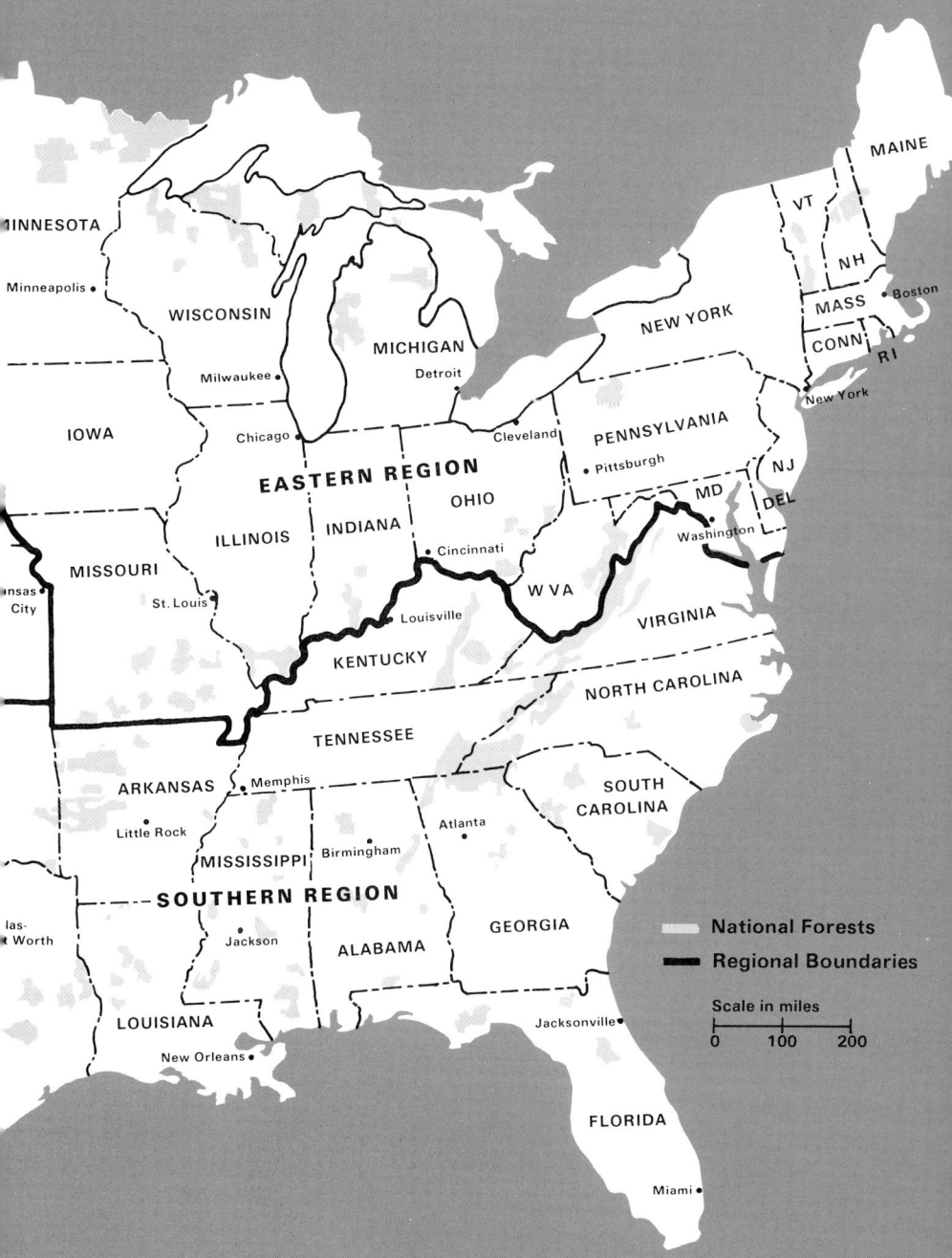

Shorn of trees, the debris consumed by fire, this is how a now-forested valley in the Jefferson National Forest near Mount Rogers, Virginia, appeared in 1942.

CHAPTER I

backyard to megalopolis

East of the 100th meridian, the line historically used to divide the dry western portion of the nation from the water-rich East, lie 50 national forests. Encompassing 23,758,013 acres[1] owned by the public and administered by the U.S. Forest Service, these national forests of the eastern United States girdle Lakes Michigan and Superior, cap the Ozark and Appalachian Mountains, and arc across the Coastal Plain and Piedmont from Texas to North Carolina.

A verdant counterpoint to the crowded urban corridors and nodes of the East, the scraps and fragments of national forests occupy almost the same amount of land area as the East's metropolitan centers. Aggregated, they amount to slightly less than the total acreage of the state of Virginia. It is perilous to generalize about such a diverse land system, but the eastern national forests appear relatively insignificant in overall national forest statistics for acreage, timber production, and recreational use. As available public land in large blocks, however, they loom large indeed; while there are 30 eastern national forests of more than 250,000 acres, in all the East there are only a dozen other public land areas of this size. And every one of the national forests of the East lies within a day's drive of a major metropolitan center. Most are even closer. National forests are the backdrop to the Atlantic Coast megalopolis, the North Woods of the Lake States, and the vacation refuge of hundreds of thousands of urbanites.

A complex interdependency, often unrecognized and unacknowledged, exists between metropolitan areas, the forests, and rural communities in the forest environs. Urban residents depend on the forests for recreation and products as diverse as baseball

I've had a "we-they" feeling, that's changed in some ways, about those who cut into forest resources. Environmentalists wore the "white hats" of the good guys. People like developers wore the "black hats." Even though I live in a house made with wood, I had not made the connection between the forests and our basic need for shelter. I once had a "park" attitude about the national forests. But now I've learned that they have a dynamic mission—multiple use, from support of wildlife and water quality to providing timber.

SALLY BATTLE League of Women Voters Columbia, South Carolina

bats, paper packaging, and lumber for homes. And in numerous ways, local economies in forest communities depend on the buying power of metropolitanites.

The 50 eastern national forests vary widely in size, land form, vegetation, mineral resources, and use. The 729,581-acre White Mountain National Forest of New Hampshire and Maine boasts peaks that rival some western mountains in scenic grandeur. But in the 163,561-acre Wayne National Forest in Ohio's Appalachia, large tracts of public land are scarred by strip mining. The Coastal Plains forests, like the 104,511-acre Oconee in Georgia and the 595,589-acre Kisatchie in Louisiana, count some of the most productive timberland in the United States.

While road maps (and, misleadingly, some Forest Service maps as well) commonly identify national forests as large, solid blocks of green, the reality is far different for the forests east of the Great Plains. The area within the forest boundary—where the Forest Service is authorized to acquire land—is usually a patchwork of public land and land still in private ownership. In many instances, the public does not own valuable minerals beneath its own land.

Unlike the forests of the West, which were created from large areas of public domain before the turn of the century, the national forests of the East for the most part have been purchased from private landowners over the past 65 years. The unavailability of large tracts of public land in the East, necessitating the often piecemeal purchase of land from private owners, has determined both the overall dimensions and the characteristic ownership pattern of eastern national forests. And this historic pattern of eastern forest acquisition also exerts a powerful influence over uses in and around the forests. "We own only about half of the land within the boundaries of the eastern national forests," Forest Service Chief John McGuire points out, "and we've made the other 50 percent very attractive for development."[2]

The national forests of the East, in the main, were assembled from land that nobody wanted. In the nineteenth and early twentieth centuries, millions of acres were shorn of their most valuable timber species, sometimes burned over or badly eroded, and then left behind by a timber industry that had exhausted the resource and moved West. Other forests, especially in the South, were created from grown-over fields of a marginal agriculture that had depleted the soil and disappeared. Most of the land purchased for the first eastern national forests in the early 1900's cost the government less than five dollars an acre. "Nearly all," says one observer, "were lands that had been abused, poorly protected, or ignored, whose owners were happy to unload on the federal government."[3]

Today, this same land has been healed and rejuvenated. Thousands of acres of fast-growing loblolly pine thrive on Louisiana's Kisatchie National Forest, abandoned by loggers in the 1920's as a "jungle of scrub oak, rattan and cat briers."[4] The Allegheny National Forest, cut over and burned at the turn of the century (and then known locally as the "Brier Patch"), now boasts stands of commercial oak, poplar, and cherry. The Monongahela, once referred to as "the Monongahela burn," is now sufficiently productive to be the focus of a national controversy over timber-harvesting practices. Scenic value has been returned to the slopes, valleys, and meadows of New Hampshire's White Mountains, once heavily and destructively logged.

Some of this rehabilitation resulted from the federal investment in replanting, fire protection, and timber-stand improvement. Some can be attributed to the remarkable, if brief, efforts of the Civilian Conservation Corps. Most was simply a function of time and nature's healing processes.

The rehabilitation of the eastern national forests ranks as one of the most remarkable conservation achievements of this century. These national forests are now a treasure store of scenic, timber, wildlife, mineral, wilderness, and recreational resources. The land that nobody wanted is back in demand. Loggers have returned to harvest the trees. Mining companies seek to discover and exploit the coal, oil, and other minerals. Some communities recognize the water-retaining value of the forested ridges, while others seek to dam up the valleys for water supply. Recreationists, far greater in number than before, compete to use it for activities as solitary as wilderness hiking and as gregarious as snowmobile rallies.

These varied uses and potentials of the forests have, of course, bred conflict. On Windfall Run in the Allegheny, oil pumped to the surface and intended for the crankcases of automobiles seeps into a trout stream. In the Ouachita National Forest in Arkansas, a developer seeks to build hunting cabins on a 160-acre tract of private land in the heart of a forest wildlife special management area. In Virginia's George Washington National Forest, owners of vacation homes on private land adjacent to the forest have sought to restrain timber cutting on the visible mountain slopes that constitute their view. At the same time, these private holdings prevent easy public access to thousands of acres of national forest land. In the Kisatchie, the use of portions of the forest as military

---

How important is national forest timber to the timber industry? We have mills here in western North Carolina that are 90 to 95 percent dependent on national forest timber. Without this source of timber, they would not be able to operate—and these mills are fairly large employers. On the average, the larger mills in the area get about 50 percent of their timber supply from the national forests; the large, permanent hardwood industry is about 50 percent dependent on national forest stumpage.

PETER R. MOUNT, Chairman, Western North Carolina Forestry Commission
Leicester, North Carolina

target ranges has left thousands of acres contaminated by unexploded shells. In the Green Mountain National Forest, ski resorts compete for Forest Service leases for steep forest slopes.

The increasing and frequently conflicting demands on these lands that nobody wanted create major challenges for management of the eastern forests. Because of their unique characteristics within the national forest system and the problems and opportunities these characteristics create, The Conservation Foundation believes that the eastern national forests need and deserve special attention—statutory recognition, policy direction, and implementing programs built around a new vision of the eastern forests' potential to serve a special role in the environmental health and economic vigor of their region and the nation.

EAST VERSUS WEST

Why this concentration on the 50 national forests east of the 100th meridian? The Conservation Foundation believes that they represent distinct and valuable national resources, which, in competition with the expansive western forests, are not receiving adequate attention and direction.

It is true that the eastern forests as a system are not dramatically different from *some* of those in the West. Closeness to urban centers? The Angeles, Los Padres, San Bernardino, and Cleveland National Forests in Southern California; Mt. Baker and Snoqualmie National Forests near Seattle; the cluster of national forests west of Denver—all are closer to metropolitan centers than are most eastern forests. And while the fragmented ownerships characteristic in the East are uncommon in western forests, a number there too are experiencing peripheral development pressures not unlike those common around and within the eastern forests.

Although it is accurate to say that land ownership patterns and proximity to metropolitan centers distinguish *most* of the eastern forests from *most* of the western forests, the real measure of difference is their relative importance in the regional resource base. In the 11 western states, individual national forests coalesce into large blocks of public land that dominate the resource base of large regions within states, and even the land base of entire states. In this 752,763,000-acre area, more than 18 percent of the land—nearly one acre in five—is national forest. (See Table 1.) Only in the states of Wyoming, New Mexico, and forest-poor Nevada is less than 15 percent of the land in national forest.

TABLE 1

A COMPARISON OF EASTERN AND WESTERN NATIONAL FORESTS

	Total Land Area[1]	Total National Forest Area[2]	Percentage of Total Land Area in National Forest	Total Forested Area[3]	Total Forested Area in National Forest[4]	Percen of To Forested in Nati Fore
Eastern States						
Alabama	32,678,000	638,457	2.0	21,770,000	631,872	2
Arkansas	33,324,000	2,460,713	7.4	18,277,000	2,440,385	13
Connecticut	3,116,000	—	0	2,186,000	—	0
Delaware	1,268,000	—	0	391,000	—	0
Florida	35,179,000	1,082,163	3.1	17,932,000	1,032,393	5
Georgia	37,295,000	845,790	2.3	25,545,000	850,858	3
Illinois	35,761,000	254,157	.7	3,789,000	237,392	6
Indiana	23,161,000	177,603	.8	3,908,000	172,839	4
Iowa	35,867,000	—	0	2,455,000	—	(
Kentucky	25,504,000	647,062	2.5	11,968,000	642,162	5
Louisiana	28,867,000	595,589	2.1	15,380,000	574,148	3
Maine	19,797,000	45,944	.2	17,748,000	49,151	
Maryland	6,369,000	—	0	2,960,000	—	(
Massachusetts	5,013,000	—	0	3,520,000	—	(
Michigan	36,492,000	2,696,631	7.4	19,273,000	2,557,398	13
Minnesota	50,745,000	2,808,975	5.5	18,984,000	2,617,965	13
Mississippi	30,290,000	1,135,984	3.8	16,913,000	1,121,324	6
Missouri	44,189,000	1,439,034	3.3	14,919,000	1,421,643	9
New Hampshire	5,781,000	683,637	11.8	5,131,000	670,648	13
New Jersey	4,820,000	—	0	2,463,000	—	(
New York	30,636,000	—	0	17,377,000	—	(
North Carolina	31,367,000	1,143,602	3.6	20,613,000	1,132,201	5
Ohio	26,251,000	163,561	.6	6,498,000	158,411	2
Pennsylvania	28,816,000	506,102	1.8	17,832,000	490,518	2
Rhode Island	671,000	—	0	433,000	—	(
South Carolina	19,366,000	607,005	3.1	12,493,000	597,915	4
Tennessee	26,474,000	618,894	2.3	13,136,000	613,646	4
Vermont	5,935,000	254,025	4.3	4,391,000	251,485	5
Virginia	25,496,000	1,596,882	6.3	16,389,000	1,590,568	9
West Virginia	15,413,000	957,538	6.2	12,172,000	948,469	7
Wisconsin	34,858,000	1,491,566	4.3	14,945,000	1,353,031	9
	740,799,000	22,850,914	3.1	361,791,000	22,156,422	6

	Total Land Area[1]	Total National Forest Area[2]	Percentage of Total Land Area in National Forest	Total Forested Area[3]	Total Forested Area in National Forest[4]	Percentage of Total Forested Area in National Forest
Western States						
Arizona	72,688,000	11,271,618	15.5	18,583,000	4,812,980	25.9
California	100,091,000	20,234,337	20.2	42,408,000	13,597,474	32.1
Colorado	66,485,000	14,364,726	21.6	22,534,000	10,987,578	48.8
Idaho	52,933,000	20,375,388	38.5	21,591,000	17,359,830	80.4
Montana	93,258,000	16,731,092	17.9	22,777,000	13,903,537	61.0
Nevada	70,264,000	5,112,567	7.3	7,660,000	2,847,700	37.2
New Mexico	77,766,000	9,219,378	11.9	18,313,000	5,725,234	31.3
Oregon	61,574,000	15,577,825	25.3	30,401,000	14,502,955	47.7
Utah	52,697,000	8,047,568	15.3	15,288,000	4,160,593	27.2
Washington	42,665,000	9,071,424	21.3	23,098,000	8,409,210	36.4
Wyoming	62,342,000	9,251,127	14.8	10,085,000	6,031,735	59.8
TOTAL	752,763,000	139,257,050	18.5	232,738,000	102,338,826	44.0

All land area totals are in acres.

[1] SOURCE: U. S. Forest Service, *Outlook for Timber, 1974*, Appendix I, Table I.

[2] Figures include established national forests and associated purchase units. The few eastern land utilization projects and national grasslands, as well as research and experimental areas, while technically part of the National Forest System, are not associated with a national forest and have been omitted.

[3] SOURCE: U. S. Forest Service, *Outlook for Timber, 1974*, Appendix I, Table I.

[4] Calculated from U. S. Forest Service data in "Timberland Use Summary," February 4, 1976.

NOTE: This analysis omits the six states bisected by the 100th meridian—North Dakota, South Dakota, Nebraska, Kansas, Oklahoma, and Texas. South Dakota and Nebraska have four national forests west of the 100th meridian, totaling 1,400,000 acres; Texas and Oklahoma have forests totaling just over 900,000 acres located east of the meridian. Because of its vast area and remoteness, Alaska has also been excluded.

In the 31 easternmost states,[5] however, national forests comprise only a small fraction—3 percent—of the total land area. Only New Hampshire has more than 10 percent of its area in national forest. For those eastern states with relatively substantial acreage in national forests, the percentage of national forest land to total state land area ranges between 5 and 8 percent (Arkansas, 7.4; Michigan, 7.4; Minnesota, 5.5; Virginia, 6.3; West Virginia, 6.2). In other eastern states, the figure is less than 5 percent, and in some of the states, less than 1 percent. (Maine, .2 percent; Indiana, .8; Illinois, .7; Ohio, .6). Eight eastern states have no national forests at all.

Even more significant is the proportion of the national forests to the total forest land base, since that is what must be considered in terms of the land available for timber and outdoor recreation in a given region. In the generally arid West, where forests do not commonly carpet the landscape as they do in the more fertile and water-rich East, national forests frequently occupy not only the *best* timber-producing land, but substantial portions of *all* the forest land. National forests in the West comprise 44 percent of all the forested land; the comparable figure for the East is 6.1 percent. In Idaho, four out of every five acres of forest land is in a national forest; in Montana and Wyoming, three acres out of five; in Colorado and Oregon, nearly one acre in two. No western state has less than 25 percent of its forest land in national forest.[6] By contrast, in only four eastern states—Arkansas, Michigan, Minnesota, and New Hampshire—do national forests comprise more than 10 percent of the state's forested land.

In the East, through the circumstances of their history, national forest landholdings are typically fragmented; the forests coexist with privately owned land which meanders through the forest valleys and occasionally onto the ridges of the Appalachians, or is interspersed in random patches in the flatter Lake States and Coastal Plains forests. In the West, the forested mountains of the national forests are distinct from the surrounding private land; in the heavily forested East, it is often difficult to tell where a national forest stops and private land begins.

The different relationships between private lands and public forests in the eastern and western regions impose different roles and responsibilities upon the public forests. Because the western forests dominate their surrounding resource base, the Forest Service must manage them to meet a substantial portion of all forest demands. But in the East, where the national forests are so small a proportion of the total resource base, private land can and must provide a larger share of certain forest activities and products. The eastern national forests may be assigned a more specialized role within their region, with management designed to meet those needs which cannot be met by private land or state and local forests and parks.

THE AMERICAN FOREST

When European colonists first came to the east coast of North America, they found a land that was almost entirely forested. In some areas, there were unbroken expanses of dense climax vegetation; elsewhere clearings stood, where wind, lightning, or the fires set by woodland Indians had temporarily removed the overstory. Large areas showed the evidence of past catastrophes—ranks of saplings growing up where the tall timber had been burned or overturned years before. Even allowing for understandable exaggeration in written reports, it is clear that early observers saw a forest much more extensive and far superior to the eastern woodlands today. A modern account describes what they encountered:

> The American hardwood forest of history—the domain of woodland Indians, the forest which was so dangerous and unlivable in the eyes of the first English settlers and which we call primeval today—was in truth a luminous, youthful, supple forest, new-born out of the Ice Age. In the nobility and quality of its trees, bushes, vines, and flowers; in the purity of lakes and streams; in the abundance and color of its birds and fish and in the personalities of its animals, no other forest that ever grew on earth could be compared with it.[7]

The early settlers were particularly impressed by the abundance of the forest's wildlife, the quality of its timber, and the purity of its waters. Many animal species then common are now extinct in the East, or survive only as scattered relic populations: Bison were found in Virginia and Pennsylvania; elk roamed in the Appalachians; wolves, bobcats, and mountain lions inhabited both the northern and the southern forests. An account of a "big hunt" or "circle drive" in central Pennsylvania in 1760 suggests the quantity and variety of the wildlife found there. The 200 participants reported killing 98 deer, 111 bison, 2 elk, 109 wolves, 41 mountain lions, and 114 bobcats.[8] The original American forest boasted huge quantities of large oaks, yellow poplar, cypress, hemlock, hickory, and the now-vanished American chestnut. So tall and straight were the white pines of New England that a number were declared property of the British Crown and marked with an axe as future masts for the Royal Navy. An early logger called the mature pine "the whales of the forest."[9] Travelers attested that almost any forest stream could be counted on as a source of wholesome water.

As the East was settled, then industrialized and urbanized, the original forest shrank and fell drastically in quality. By the end of the 1800's, the once-great forests of the Northeast were gone, their timber cut to build the great cities of a maturing nation. In some areas, such as Western Pennsylvania, the forests had been devastated; elsewhere, the most desirable species—especially white pine —were severely depleted. Gifford Pinchot, the founder of the Forest Service and giant of American forestry, wrote that when, as a young man, he returned to the United States in 1890 from his forestry studies in Europe," ... the most rapid and extensive forest destruction ever known was in full swing."[10] With eastern timber exhausted, the timber companies began to carve into the rich forests of the Northwest.

Those who condemn the old logging industry do not take into account the environment in which people lived in those times. Cutting the timber was a means of livelihood just like growing wheat in a prairie state. It was natural that people who lived in the forest looked to natural resources for their sustenance and their livelihoods. Today, critics of their methods are unsympathetic to the things a pioneer had to do in order to survive. The use of timber was as much his livelihood as transactions in stocks and bonds on Wall Street are to a stock broker.

SHERMAN ADAMS, Owner, Loon Mountain Ski Resort
Former Governor, State of New Hampshire
Lincoln, New Hampshire

THE EVOLUTION OF NATIONAL FORESTS

The threat of continued forest devastation made America's forest resources a matter of national concern. At first, interest focused on the virtually unrestrained cutting of timber on the expansive public lands of the West, where large commercial operators adroitly evaded both the law and prosecution. By subterfuge, they acquired rights to vast amounts of public timber. In an attempt to protect some remaining public land from exploitation, the Congress in 1891 enacted legislation permitting the President to set aside portions of the public lands as "public reservations" in which the land and resources would be retained in public ownership.[11] The legislation permitting the reservation of land was the mechanism for the creation of the first "forest reserves," the forerunners of the present national forests.

In 1897, Congress enacted the Organic Administration Act[12] establishing the fundamental direction for management of the forest reserves. At the time, Congress was apprehensive that excessively large areas of public domain might be withdrawn from private exploitation, so the legislative draftsmen framed the act in the negative:

> No national forest shall be established except to improve and protect the forest within the [national forest] boundaries or for the purpose of securing favorable conditions of water flow, and to furnish a continuous supply of timber for the use and necessities of citizens of the United States.

With the authorization these acts provided for the Federal Government to own vast tracts of forest land, concern mounted that these properties, because they would never be in private ownership, would not provide tax revenue for state and local governments. In 1906 Congress enacted legislation requiring the Federal Government to return to the states containing national forests 10 percent (increased to 25 percent in 1908) of the revenue collected from the forests.[13] The revenues, raised primarily from timber sales and leasing of federally owned minerals, were to be used for schools and roads in the counties containing forests.

The problems of New England are different from those in the huge national forests in the West. When Congress enacts laws, they enact them for the entire country. Some of the towns and counties out West, particularly where you have huge marketable timber, receive a very high return per acre from the national forests. We receive practically nothing.
RICHARD CLARK, Selectman
Ripton, Vermont

In 1898, soon after passage of the Organic Administration Act, Gifford Pinchot was appointed director of the Division of Forestry (later to become the Bureau of Forestry) in the Department of Agriculture. By Pinchot's own claim, he was the "first American to make forestry his profession."[14] Possessing a powerful personality, political shrewdness, and the ability to inspire his subordinates, Pinchot presided over the transformation of the anemic Division of Forestry into the Forest Service. In the process, he established the tradition of professionalism and esprit de corps that endures today. He also set forth fundamental objectives for the forests themselves in the preface of a slim staff manual known as the *Use Book:*

> Timber, water, pasture, mineral and other resources of the forest reserves are for the use of the people. They may be obtained under reasonable conditions without delay. Legitimate improvements and business enterprises will be encouraged.[15]

The division Pinchot took over had been established in 1881 to conduct experimental research and studies of forest resources, primarily for the Department of Interior's General Land Office, which had authority over the vast public domain and forest reserves in the West. The Division of Forestry, with but 11 employees, had no forest management authority. It was, in Pinchot's words, "a microscopic and mixed outfit—some good and some not so good."[16]

Foreclosed from direct work on the federal forests, Pinchot concentrated his division's efforts on assistance to private landowners. (This continues—with the cooperation of state forestry agencies—to be a major responsibility of the Forest Service's State and Private Forestry branch.) However, his real objective was to secure the division's control over the federal forests. He wrote later: "To get charge of them [had] become my chief objective in life."[17] Pinchot had fostered a friendship with conservation-minded Theodore Roosevelt and encouraged the President's interest in forest problems. Roosevelt's first State of the Union message in 1901, in a section drafted by Pinchot and two associates, asserted that "forest and water problems are perhaps the most vital internal questions of the United States."[18]

Roosevelt went on to describe the dispersed responsibility for forest resources—protection under the Department of Interior's General Land Office, mapping under the Geological Survey, the preparation of working plans under the Bureau of Forestry (the division had been upgraded to bureau status just a few months earlier)—and concluded, "These various functions should be united in the Bureau of Forestry, to which they properly belong."[19]

Because western interests were more than satisfied with the unaggressive administration of forest lands by the Department of Interior's General Land Office, four years passed before the political climate favored Pinchot, thanks in large part to the work of the American Forestry Association and other citizen interest groups. On February 1, 1905, Roosevelt signed the Transfer Act, which placed management of the federal forest reserves in the eager hands of Pinchot and the Bureau of Forestry. Of less significance, but highly gratifying to Pinchot, was the fact that the Agriculture Appropriations Act, approved by the Congress four months later, referred not to the Bureau of Forestry but to the "Forest Service" as the custodians of the federal forests.

ESTABLISHING THE EASTERN NATIONAL FORESTS

While the legislation of the 1890's authorized the creation of national forests (the name was changed from "reserves" in 1907) from land in the public domain, this primarily affected the West, for by then little public land remained in the heavily settled East. In 1900, to assess the need for public land in the East, the Congress directed the Secretary of Agriculture to "investigate the forest conditions in the Southern Appalachian Mountain region of western North Carolina and adjacent states."[20] In January 1901, Secretary of Agriculture James Wilson sent his report to President William McKinley. Wilson's letter of transmittal described the situation:

> The rapid consumption of our timber supplies, the extensive destruction of our forests by fire, and the resulting increase in the irregularity of the flow of water in important streams have served to develop among the people of this country an interest in forest problems which is one of the marked features of the close of the century. In response to this growing interest the government set aside in the western forest reserves an area of more than 70,000 square miles. There is not a single forest reserve in the East.[21]

Not surprisingly, given his position as Secretary of Agriculture and the influence of Pinchot, Wilson recommended the establishment of a forest reserve in the Southern Appalachians, an area that had been proposed for a new national park. But at the turn of the century, no authority existed for the Federal Government to buy land for forests. Legislation authorizing purchase funds for forest reserves was introduced that year but failed to pass then and in several subsequent Congresses.

However, support for the concept of forest reserves in the East was growing. From the Southern Appalachians it spread to New England, where it was promoted by the Society for the Protection of New Hampshire Forests and others, to the Ozarks, the Hudson Highlands, the headwaters of the Mississippi, and to Texas.[22] Pinchot wrote later: "It was this combined pressure that finally overcame the resistance of the [House] Rules Committee and that

If you are going to see a significant expansion of the national forest system, there has to be more than just the hikers or fishermen or snowmobilers behind it. Many people and many interests have to be involved, because that is what the politicians will listen to.

If it had just been recreational interests behind the White Mountain National Forest, the forest never would have happened. It had—and has—to be more than that. The Society's interest was protection of the forest—for economic reasons as well as its recreational values. Those who worked for establishment of the White Mountain equated protection of that forested land with the economic vitality of the region, particularly the recreation industry—hotels and the railroads that brought people to those hotels. Because they made national forest establishment an economic issue rather than just recreation or wildlife or aesthetics, they were very successful. If you define the public interest broadly, you come up with solid economic justification for forest acquisition.

PAUL O. BOFINGER, Executive Director, Society for the Protection of New Hampshire Forests, Concord, New Hampshire

---

of that famous idealist, Joe Cannon, Speaker of the House, whose position was 'Not one cent for scenery.' "[23]

Costly and tragic floods, like that of the Monongahela River in 1907, also generated support for national forests in the East. Writes C. R. McKim in his *Fifty Year History of the Monongahela*:

> West Virginia was rich with the natural resources other parts of the country needed to develop their own industries. In what is now the Monongahela grew some of the greatest stands of hardwood timber to be found in any part of the world, and it was reaped by numerous logging operations. It was there for the taking with few restraints. Exploitation was the order of the day . . .[24]

But depletion of the timber resources created even more disastrous consequences:

> During March 1907, heavy rains brought flood waters down the Monongahela River . . . the trees and other healthy vegetation were no longer there to regulate the rainwater's flow. It devastated all the rich

agricultural land in the basin of the Monongahela River, causing some $100 million in damages—a gigantic sum for those times—then descended in all its fury upon the helpless city of Pittsburgh, causing there additional damages of $8 million, drowning people and ruining their homes.[25]

Following this disaster, to prod Congress to take action, the West Virginia legislature in 1909 enacted legislation permitting the United States to buy land for what became the Monongahela National Forest, more than two years before Congress passed the Weeks Law.[26]

The Weeks Law,[27] enacted in 1911, was the progeny of the forest-purchase legislation first introduced in 1901. It provided for the purchase of "forested, cut-over or denuded lands within the watersheds of navigable streams . . ." The emphasis on protecting stream flow was deliberate, intended to avoid the issue of unconstitutionality by linking the acquisition of forest land to the constitutional authority of the Federal Government to regulate commerce. It was not until 1924 that the Clarke-McNary Act added "the production of timber" as a purpose for forest acquisition.

The immediate objective of the Weeks Law was the purchase of five million acres of forest land in the Southern Appalachians and another million acres in the White Mountains in New Hampshire.[28] It carried an appropriation of $9 million to be spent over six years for forest acquisition in those mountain regions. However, eastern national forest aspirations quickly expanded. A 1914 Forest Service study urged the establishment of a national forest in Missouri.[29]

In 1916, sufficient land had been acquired to establish the first eastern forest, Pisgah National Forest in North Carolina.[30] In 1918 the Pisgah was joined by three more—the Shenandoah in Virginia (now a part of the George Washington); the Natural Bridge, also in Virginia (now a part of the Thomas Jefferson); and the White Mountain. (The Alabama—now the William B. Bankhead—also was established from public domain land in that year.)

In 1920 five more Appalachian forests were created—the Boone, now part of the Pisgah, and the Nantahala in North Carolina, South Carolina, and Georgia; the Cherokee in Tennessee; the Unaka in North Carolina, Tennessee, and Virginia; and the Monongahela in West Virginia. The Allegheny National Forest in Pennsylvania was established in 1922. Through the remainer of the 1920's, acquisition focused on filling in the established forests.

I fully realize that when the Allegheny National Forest was created, there was nothing here. Most of this land was nonproductive because it had been logged off. It was worthless. Everybody was exploiting everything. The Allegheny National Forest took over this land and brought it up to the point where they could make one or two million dollars off the acreage during the course of a year. I think this is good. It showed the way to private landowners. However, I think the government has enough land. This country has been based on private ownership, and it should continue to be based on private ownership.
DAVID K. RICE, Chairman, Board of Commissioners
Warren, Pennsylvania

The Great Depression stimulated a boom decade of national forest establishment and land acquisition, combining resource protection with national socio-economic objectives. Land abandoned by owners who could not pay the taxes was acquired by the government very cheaply. Local people were desperate for any activity that would pump money into a community, so they welcomed establishment of forests which provided for federal investment in otherwise unused land and generated badly needed jobs. And national forests provided a work place for President Roosevelt's Civilian Conservation Corps.

During the 1930's, some 26 national forests were established, ranging from the Clark and Mark Twain on the Ozark Plateau of Missouri [31] to the Green Mountain in Vermont; from the Chequamegon and Nicolet in Wisconsin to the Osceola and Apalachicola in Florida. In Missouri, the Depression helped persuade the state legislature to remove restrictions on federal acquisition, thus permitting the establishment of the state's two large national forests.[32]

Since World War II, only three new forests have been established outright—the Wayne in Ohio, the Hoosier in Indiana, and the Uwharrie in North Carolina. A number of small forests in the South also were established, using land acquired by the Federal Government during the Depression.

As it assembled its new lands in the East, the Forest Service concentrated on acquiring the surface rights, which are traditionally bought and sold separately from the subsurface rights in areas likely to contain mineral deposits. Because the purchase price of mineral rights below the surface was prohibitive, the Forest Service often left these in the hands of other owners. Today, the government owns only two-thirds of all the mineral rights under its eastern forests, and a much smaller percentage in mineral-rich areas.

FOREST USES

That through circumstances of personality and historical accident the national forests came to be administered by an agency of the Department of Agriculture and not by the Department of Interior largely determines the purposes and uses of these public lands. The national parks are administered as outdoor museums to be preserved and enjoyed, while the national forests have been, since their inception, working resource lands. When opting in 1901 for purchase of land in the Southern Appalachians for forest reserves rather than a national park, Secretary of Agriculture Wilson used adroit wordsmanship to justify his decision: "The idea of a national park is conservation, not use; that of a forest reserve, conservation by use."[33]

I wouldn't like to see the White Mountain National Forest turned into a park, although some of the more intensively used areas have to be managed like a park. The different uses of the forest are compatible with our needs around here. Many companies depend on the national forest for timber. We like to enjoy the forest when people are not here. But you have to remember when cars are jammed up in town on a summer day, that the economy of the town depends upon those cars.
JUDITH SMITH, President, American Association of University Women
North Conway, New Hampshire

Indeed, without the prospect of immediate or future use, rather than preservation, it is unlikely that 24,000,000 acres of public forest land would exist today in the East. Pinchot stressed use in lobbying for national forests. Their acquisition was predicated on the understanding that Americans would receive benefits—usually tangible economic benefits—from them.

While nature is generally allowed to take her course in the national parks, the national forests are managed for five uses. As officially stated in the Multiple-Use Sustained-Yield Act of 1960: "It is the policy of the Congress that the national forests are admin-

> *Multiple use is the best use of the forest for all the people. Under multiple use, you get much greater use of the forest than, for example, under the "forever wild" concept of the Adirondack Park in New York State.*
> RICHARD H. BURT, Vice President/Works Manager, Allen-Rogers Corporation
> Laconia, New Hampshire

istered for outdoor recreation, range, timber, watershed, and wildlife and fish purposes."[34] Other laws provide for the mining of government-owned minerals on national forest land.

But the national forests have served purposes beyond those listed in the statutes. During the Great Depression, they were important parts of the engine of economic recovery. During World War II, an entire 350,000-acre national forest in Florida was transformed into what is now Eglin Air Force Base. Today, uses of national forest land range from the provision of waste-disposal sites to military bombing ranges.

Gifford Pinchot and his associates envisioned the national forests as exemplary demonstrations of the best in forestry and conservation techniques. Under wise management, the woodlands would be returned to high productivity, furnishing the nation with a continuous source of timber as well as flood protection, water supply, and wildlife. But at the turn of the century, neither Pinchot nor other national forest advocates could have imagined the popularity the eastern national forests would gain, the uses to which they would be put, or the demands that would be made upon them three-quarters of a century later.

Those instrumental in the establishment of the first eastern national forests had a vision of restored forests perpetually contributing to the basic needs of a developing nation. That vision has largely been fulfilled. But society and its needs have changed dramatically since these forests were established, creating new and increasing demands upon the forests. The Conservation Foundation believes that present laws, institutions, and processes cannot adequately cope with the new requirements for resource allocation implicit in the increasing demands. Choosing among these competing demands requires a new vision for the eastern national forests, a long-term vision that illuminates the special ways these public forests, as an element of the total resource base east of the 100th meridian, can help fulfill society's future needs.

Such a governing vision and effective management policies to implement it must be developed, for uses of the forests and changes and activities on private land around them reduce future options daily. Paving roads and parking lots on privately owned parcels reduces the watershed benefits of public lands. Development causes pollution and siltation of streams, impairing their value for water supply, fishing, and swimming. Intensive drilling for oil and strip mining for coal reduce the value of the surface for both timber production and recreation. Use of the forest land for utility corridors and other urban support facilities further degrades the total forest environment. Some of these uses may be necessary and appropriate. But decisions are often made ad hoc. Policy tends to be based on a short-term view rather than a long-range vision of how the forests should serve future generations.

People see the fights over clearcutting and wilderness designation on the national forests as symptomatic of the failure to establish a national policy based on clearly established goals—in other words, an idea of what the forests are to be, what purposes they are to serve.
LUCY SMETHURST, Chairman of the Board, The Georgia Conservancy
Atlanta, Georgia

The eastern national forests' fragmented land-ownership pattern and proximity to metropolitan areas necessarily affect future opportunities and management policies for these lands. These characteristics pose other challenges to effective management:

- The national forests and adjacent private lands constitute complex natural resource systems with complex interactions between water quality and quantity, timber production, wildlife habitat, and the type and quality of recreational opportunity.

- The interdependent land and water resources on public and private lands are planned for and managed by a variety of authorities—federal, state, and local governments; multi-state regional commissions; and multi-county development districts. While authority is guarded jealously, actual responsibility is diffused and confused.

- While the national forest land base in the East is relatively small, it is in high demand for timber production, mineral extraction, wilderness and watershed protection, and recreational uses. Competition among users for this limited resource base can be severe. In some cases, there is a direct conflict between users; in other instances, it is difficult to achieve a balance of uses satisfactory to everyone.

Of course, funding and political realities also constrain forest management options. In the face of calls for budget restraint, the Forest Service must compete with other federal agencies for limited funds. And within the Forest Service itself, there is intense competition for available funds among the various functional activities (e.g., timber management, recreation, wildlife). At another level, national forest regions and individual forests within those regions compete with one another for funds.

Political factors circumscribing activities of the Forest Service include pressures affecting congressional authorization and appropriations for Forest Service activities; pressure exerted by state and local governments in support of a favored project (such as a dam, highway, or ski complex) or in opposition to unpopular proposals (such as wilderness designation or additional national forest acquisition); or pressure from timber interests to increase allowable cuts on the national forests.

Within these constraints, forest management can be designed to meet more effectively the needs and demands of Americans at the threshold of the twenty-first century. Some current trends suggest appropriate roles for the eastern national forests. For instance, as urban centers and suburban areas increasingly blur into a congested east-coast megalopolis, the forests can help shape future urban development and preserve open space, enhancing the environmental quality of the entire region. They can provide a range of outdoor recreational opportunities to fill the increasing leisure time of city dwellers anxious to escape congestion. Higher gasoline prices and potential fuel shortages make these forests, so close to major population centers, particularly attractive.

Increasingly, individuals seem to be seeking more personal autonomy and more satisfying life styles, often through less regimented and conventional forms of work and recreation. Many find this new life in rural areas of moderate climate and scenic amenity. The eastern national forests can help expand the rural economic base through increases in timber production and recreation use, as well as by providing resources for labor-intensive, low-technology enterprises.

There used to be a Job Corps camp here in Ripton. The Black and Puerto Rican kids were up there learning machine operation, timber cutting, and all kinds of things. Those Forest Service guys threw their hearts into working with those kids.

The Forest Service is trying to find mechanisms by which concerns of the forest and wildlife preservation can relate to the problems of the inner cities. How do you go into Boston and create interest among Blacks in matters relating to forests and trees? I think we are developing, among other things, two Americas—one is the Detroits and the other is the rest of us. I feel this is a real frontier-type issue which must be addressed. There is a real fear that those of us in the environmental movement will lapse into upper-middle-class comfort.

BRENDAN WHITTAKER, Chief, Information and Education
Agency of Environmental Conservation
Montpelier, Vermont

Society's changing needs and life styles will unquestionably continue to exert new pressures upon the national forests of the East. These are evident today in conflicts between users; they can only increase as pressures for more new and intense uses of the forests mount. These changes offer opportunities, as well as challenges, for forest management.

REFERENCES

Chapter I

1. Data on acreage for the national forests are from U.S. Forest Service, *National Forest System, Areas as of June 30, 1975* (Washington: GPO, 1976).
2. Interview with John R. McGuire, Chief of the U.S. Forest Service, June 19, 1974.
3. Glen O. Robinson, *The Forest Service: A Study in Public Land Management* (Baltimore: Johns Hopkins Press for Resources for the Future, 1975), p. 11.
4. Michael Frome, *Whose Woods These Are: The Story of the National Forests* (Garden City, N.Y.: Doubleday and Co., Inc., 1962), p. 135.
5. Being neither truly east nor west, the six states bisected by the 100th meridian —North Dakota, South Dakota, Kansas, Nebraska, Oklahoma, and Texas— have been omitted from the calculation. Of the six, Texas and Oklahoma have national forests east of the meridian, South Dakota and Nebraska have national forests west of the meridian.
6. There is one meridian-state anomaly worth noting: In South Dakota, the Black Hills National Forest contains 80.9 percent of all the forested land in that state.
7. Rutherford Platt, *The Great American Forest* (Inglewood Cliffs, N.J.: Prentice Hall, 1965), p. 43.
8. Victor E. Shelford, *The Ecology of North America* (Urbana: University of Illinois Press, 1963), p. 28.
9. John Springer, *Forest Life and Forest Trees*, quoted in Ogden Tanner, *New England Wilds* (New York: Time-Life Books, 1974), p. 29.
10. Gifford Pinchot, *Breaking New Ground* (N.Y., N.Y.: Harcourt Brace and Company, 1947), p. 23.
11. Creative Act of 1891 (26 Stat. 1103; 16 U.S.C. 471).
12. Organic Administration Act of 1897, as amended (16 U.S.C. 473-478, 479-482, 551).
13. EBS Management Consultants, Inc., *Revenue Sharing and Payments in Lieu of Taxes on the Public Lands*, Volume II, a report prepared for the Public Land Law Review Commission (n.p.: 1970), p. 17.
14. Pinchot, *Breaking New Ground*, p. 30.
15. U.S. Forest Service, *The Use of the National Forest Preserves*, quoted in Gifford Pinchot, *Breaking New Ground*, p. 266.
16. Pinchot, *Breaking New Ground*, p. 137.
17. Ibid., p. 244.
18. Ibid., p. 191.
19. Ibid., p. 190.
20. U.S. Department of Agriculture, *Forests, Rivers, and Mountains of the Southern Appalachian Region* (Washington: GPO, 1901), p. 166. (Hereafter cited as *Forests*.)
21. Ibid,. p. 168.
22. John Ise, *United States Forest Policy* (New Haven, Conn.: Yale University Press, 1920), p. 211.
23. Pinchot, *Breaking New Ground*, p. 240.

24. C. R. McKim, *Fifty-Year History of the Monongahela National Forest* (U.S. Forest Service, n.d.), p. 3.
25. Ibid., p. 3.
26. Ibid., p. 4.
27. Weeks Law, Act of March 1, 1911, as amended (16 U.S.C. 480, 500, 513-517, 517a, 518, 519, 521, 552, 563).
28. U.S. Congress, *A National Plan for American Forestry* (Washington, D.C.: Government Printing Office, 1930), p. 1172.
29. U.S. Forest Service, "Open Space for People," 1973, p. 6.
30. Data on dates of forest establishment are from U.S. Department of Agriculture, *Establishment and Modification of National Forest Boundaries: A Chronological Record, 1891-1973* (Washington, D.C.: U.S. Forest Service), 1973.
31. These two forests were consolidated as the Mark Twain National Forest in 1976.
32. U.S. Forest Service, *Plan For Managing the National Forests in Missouri* (n.p., 1976), p. 3.
33. U.S. Department of Agriculture, *Forests*, p. 167.
34. Multiple-Use Sustained-Yield Act, Act of June 12, 1960, as amended (16 U.S.C. 528-531).

Logging train hauls timber from Pennsylvania's Allegheny National Forest in the 1930's.

CHAPTER II

present uses and future demands

When the management of the national forests was turned over to the newly named Forest Service in 1905, Gifford Pinchot articulated its cardinal principle of use as well as protection of the lands in its care:

> All the resources of forest reserves are for use, and this use must be brought about in a thoroughly prompt and businesslike manner, under such restrictions only as will insure the permanence of these resources.[1]

The Multiple-Use Sustained-Yield Act of 1960 made this governing management principle an official statutory charge in defining the use categories of "outdoor recreation, range, timber, watershed and wildlife and fish." By "multiple use," the law does not mean that the Forest Service must accommodate each of the uses on every acre under its jurisdiction. It does mean that all uses generally are accounted for in plans for large areas of land. It also means managing in order to minimize conflicts among uses—designing timber cuts so they do not conflict with watershed values, wildlife, or recreation; and keeping recreational vehicles from damaging soils, for example. Indeed, there are many cases in which one use, such as cutting a mature timber stand, actually benefits another use (in this case, some forms of wildlife). Other uses—such as timber production and wilderness, or wildlife and campgrounds—are not nearly so compatible. Thus, the Forest Service, while making multiple use of the entire forest, devotes individual tracts to those uses that can reasonably coexist.

How are the forests used today? Relative demands for the various forest resources have changed in the years since they were established. How will such factors as national trends in popula-

Most people who use the forest see it with tunnel vision. They either cut the timber and don't do anything else, or hike and don't do anything else, or whatever. As a hiker, I used to be adamantly opposed to clearcutting. But put into the perspective of what the forest is all about, the cutting is no longer offensive. I think that if people could be shown how the different uses of the forest relate, they would become more tolerant of the other uses.

If, for instance, people could see how timber management can be beneficial to deer, it would help them to understand that both have their place in the forest. One of the most meaningful experiences I've had was going up to the Kilkenny unit of the White Mountain Forest and seeing how the deer co-exist with the heavy machinery used in timber harvesting, how timber harvesting can be done in conjunction with management of deer habitat. The two are compatible. It's something not many people get an opportunity to see.

Large areas of wilderness are fine, but it means that timber cutting will be increased somewhere else. That made me understand the problems facing the managers of the forest; they have to balance everything out.

JUDITH SMITH, President, American Association of University Women
North Conway, New Hampshire

---

tion, incomes, relative commodity prices, and public tastes and preferences affect demands on the forests of the East in the decades to come? A consideration of current patterns of use and probable future demands on the forests reveals present and potential conflicts among the multiple uses. Such an analysis also points the way toward the development of effective and realistic goals and objectives for the eastern national forests.

TIMBER

Of the nearly 24,000,000 acres of national forest land in the East, slightly over 21,000,000 are classified as "commercial" forest.[2] This classification means that the land is capable of growing at least 20 cubic feet of timber per acre yearly—not particularly high as forest growth goes—and has not been set aside from timber production for wilderness or other nontimber use. About 40 percent of this land now supports trees of sawtimber size—that is, trees large enough to be usable as lumber. In the northern forests, these stands are mainly hardwood types—maple-beech-birch or oak-hickory. In the South, there is an approximately equal division between the various species of pine (loblolly, longleaf, shortleaf, slash pine) and the hardwoods (oak-hickory), with considerable land devoted to mixtures of hardwoods and pines.[3] Major end products of hardwoods found on eastern national forests include furniture, pallets,[4] flooring, and railroad ties, while softwoods go principally into construction lumber, plywood, and paper.

Altogether, the eastern national forests contain about 2 percent of the nation's stock of softwoods and close to 6 percent of its hardwoods.[5] Mainly because its southern pine forests are young and fast growing, however, the eastern national forests account for nearly 5 percent of the nation's annual growth of softwood, generally considered to be in shorter supply than hardwood.[6] Each year, between 1 and 2 percent of the total wood fiber cut in the United States comes from an eastern national forest, with softwoods accounting for a somewhat larger share of the cut than hardwoods.[7] While the eastern national forests clearly are not a major factor in national wood production, timber interests are quick to point out that their products can be important to local economies.

Demand is increasing for this federally owned timber, although in only a few eastern forests do Forest Service timber managers find the kind of keen market interest that exists for the pines and firs of the Pacific Northwest and California. At present, the eastern national forests are growing about three times as much softwood and about five times as much hardwood as is being removed.[8] Much of this wood is on small trees, not yet grown large enough to harvest. Some of these trees, particularly the southern pines and other softwoods, eventually will have significant commercial value. Others, both large trees and immature ones, have little market value, for they are "inferior" species, or trees too bent or crooked to make good lumber.

The Forest Service sells logging contractors the right to cut timber by competitive bid. From 1973 to 1975, it raised an average of $33 million yearly from sales on eastern national forests, a fairly modest amount compared with its receipts in the West ($371 million).[9] These timber revenues constitute by far the largest income from these forests, amounting to many times the revenues raised from campground fees, grazing fees, and special-use permits.[10] It is not clear, however, whether the government actually makes much of a profit from timber management on its eastern forests. Direct costs of marking sale sites and administering sales offset some $14 million in revenue (three-year average).[11] If no other costs are allocated to timber production, the Forest Service could claim a yearly net return from eastern national forest timber of $19 million, or about 95 cents per acre from its 21,000,000 acres of commercial forest land.

These direct costs, however, are by no means all of those which should be charged against timber production. The cut trees embody past investments in reforestation, fire protection, and timber-stand improvement, as well as the foregone interest on this investment between the time it was made—sometimes as much as 40 years ago—and the time when the trees are harvested.[12] The precise value of this investment component is unclear; unfortunately, the Forest Service does not cumulate these costs with respect to the trees it sells annually. Moreover, the Forest Service currently spends an average of $20 million a year on forest roads and trails, some of which are constructed primarily for timber management. Thus, although data limitations prevent the determination of an exact figure, it seems evident that the government's annual "profit" from its eastern national forest timber operations is considerably below 95 cents per acre.

Since the early 1960's, most of the eastern national forests have used an even-aged system of timber management, which the Forest Service considers the most efficient system available. Under this system, timber is cut in such a way that the forest is eventually composed of a series of stands of trees, within each of which all of the trees are approximately the same age. Over the entire forest, there are thousands of these even-aged stands (which might range from 10 to 100 acres in size), the age of each stand varying with the date it was last cut. With a 70-year "rotation," for example, one-seventieth of the stands would be one year old; one-seventieth would be two years old, and so on. Producing a fully regulated forest takes generations, of course, and most of the eastern national forests are far from having a uniform distribution of age classes.

In order to produce even-aged stands, the Forest Service requires timber operators to cut most or all of the trees on a given site, either clearcutting it or—less frequently—leaving a few trees to seed or shelter the parcel's future growth.

Data on the degree to which the Forest Service uses the various types of timber management are not good. The best statistics available indicate that, as late as 1973, less than 5 percent of the area regenerated in regulated harvest in eastern national forests was cut using the "selection" system, which is aimed at creating or maintaining an uneven-age composition.[13] But the predominance of

A recent clearcut in Florida's Ocala National Forest.

Longleaf pine regenerating on another Ocala clearcut.

clearcutting as a management technique may diminish as a result of an August 1975 decision by the Fourth Circuit U.S. Court of Appeals, which held that clearcutting on West Virginia's Monongahela National Forest violated the Forest Service's Organic Act. Reacting to this, the Congress passed the National Forest Management Act of 1976,[14] directing the Secretary of Agriculture to draw up new guidelines for timber management in all national forests.

The Forest Service does not manage all of its commercial timberland for timber. Some of these areas are not attractive for timber production because they are too steep, too inaccessible, too far from lumber and pulp mills, or stocked with undesirable tree species. Cutting is not allowed in other areas because they border roads or scenic rivers, or contain historical or wildlife resources that would be disturbed by timber cutting. Along travel corridors, logging may be prohibited within a certain distance from the roadway. Within a somewhat wider corridor, there are often restrictions against piling cutting debris high enough to be visible from the roadway. Still other areas can be cut only when special care is taken in the manner or intensity of the cut. Much of eastern land cut by the selection method is within such special-treatment zones.

At any one time, the amount of land disturbed by recent cutting is relatively small. From 1968 to 1973, some 2,300,000 acres of eastern national forest land experienced cutting—about half in final harvest, mainly clearcut, and the rest in an intermediate harvest or thinning. The area cut was about 13 percent of the federal land in the southern (Region 8) forests; about 6 percent of the land in the northern ones.[13] If it takes roughly 25 years for regeneration to remove most evidence of a clearcut, perhaps a fifth of the forest at any one time shows evidence of timber activity.[15] (After 10 years or so, the main evidence is obviously young stands of trees.)

Principally to gain access to this timber and to protect it from fire, the Forest Service maintains nearly 30,000 miles of roads, mainly dirt or gravel, in the eastern forests. Roads are also helpful to recreationists, hunters, and other forest users, but they are indispensable to the production of timber. Many new roads, in fact, are built by timber purchasers, who are given allowances for their cost by the government through adjustments in the timber-sale price. These roads have the side effect of opening up motorized access to the forest for other users. They also tend to reduce the size of potential wilderness areas, interfere with solitude-loving wildlife, and cause a certain amount of erosion.

Timber management on the eastern national forests differs in several ways from that practiced on private lands. Primarily, this is because, by law, the Forest Service must manage forests for "multiple uses" and for a "sustained yield" of products—goals that may or may not be followed by private or industrial landowners. "Sustained yield" is as ambiguous a term as "multiple use." Legally, it means "the achievement and maintenance in perpetuity of a high-level annual or regular periodic output of the various renewable resources . . . without impairment of the productivity of the land." This has been variously interpreted to mean an even annual flow of products or wide swings in production levels, so long as the land's ability to produce is maintained. Regardless of the interpretation, both requirements impose a duty on the Forest Service that is not borne—except voluntarily—by private landowners. Several additional Forest Service management practices are related to its legislative mandates, but not automatically implied by it.

First, the amount of national forest timber offered for sale is much less responsive to short-run changes in demand than timber supplied by private owners. Allowable sales volumes for a given forest usually are set for a 10-year period. Stumpage buyers have some leeway to vary the rate at which they cut their purchases (contracts generally allow them three years to cut), but there is relatively little short-run variation in the volume of stumpage that the government offers for sale.

Second, national forest lands are managed primarily for lumber or sawtimber-sized trees, with smaller-diameter pulpwood and poletimber sold mainly as by-products of sawtimber management.[16] This priority determines rotation ages that are somewhat longer than those of industrial woodlands. In the South, for example, private landowners cut pines in 20-to-30-year rotations for pulp; 40 to 50 years for sawtimber. The national forests in that region commonly manage for 60-to-80-year sawtimber rotations, producing larger sawlogs of the same species. Hardwood rotations in the eastern national forests may be 80 years or more, resulting in premium-quality sawlogs. In general, private forests—when managed at all—tend to be cut early, maximizing net financial return but not individual tree growth. National forest managers are less concerned than are private sector managers with the financial costs of carrying inventories of mature, slower-growing trees. Thus they can use a longer, more "biological" rotation, cutting when trees are closer to physical maturity.

A tree is felled in the White Mountain National Forest in New Hampshire. The Forest Service sells the public timber by competitive bid. Buyers may be large timber companies, but often are small timber operators who cut the trees and sell to area mills.

A skidder, its wheels girded with chains, hauls freshly cut logs from where they fell.

At a central point, the logs are cut into manageable lengths for loading onto a truck.

The truck transports the logs to a mill. The Forest Service encourages logging during the winter to minimize soil damage, erosion, and interference with recreation.

Like giant jackstraws, pine logs are stacked at a pulp mill near the Ocala National Forest in central Florida.

Third, eastern national forests are not necessarily managed for the most economically valuable tree species, for the Forest Service must balance its multiple objectives. On the one hand, it wishes to grow economically attractive species, both to increase timber revenues for the U.S. Treasury and to set an example for private landowners. On the other hand, the species most valuable for timber may not be those best suited to wildlife or recreation uses. Questions about conversion of sites from hardwoods to softwoods, mainly in the South, reveal this conflict.

After the timber companies' rapid cutting of the southern pine forests around the turn of the century, large areas had regenerated only in scrub and low-value hardwoods.[17] When the Forest Service acquired its southern lands, it emphasized the planting of fast-growing, higher-value pine species, both on barren sites and abandoned farm fields and as replacements for some of the poorer hardwoods. Considerable concern has since been expressed that this policy will reduce wildlife populations; deer and squirrel prefer hardwood or mixed-species forests. Carried to an extreme, it could produce a farmlike forest of a single tree species, visually monotonous and susceptible to catastrophic disease.[18]

Forest Service policy on such stand conversion seems to vary from forest to forest. A ranger on Florida's Appalachicola National Forest said: "We went into stand conversion a bit a few years back, but now we replant in the trees that were there before."[19] But as recently as 1975, Arkansas environmentalists claimed that national forests in that state were being converted from hardwood to pine at a rapid rate.[20] The Forest Service Manual could be interpreted to support either policy, for it calls for timber-management policy to "include the production of premium size and quality hardwoods *on the better suitable sites* [emphasis added] to alleviate the projected decline in this class of material."[21]

The future demand for any timber—whether from the eastern national forests, other national forests, or privately owned timberlands—depends primarily on three factors: the demand for wood products, the availability and price of substitute materials, and the price of wood.

The demand for timber is derived from the demand for wood products. About half of the usable timber taken out of forests today is sawed into lumber. Another third is pulped, with the remainder used for plywood, veneer, poles, and other products. Relatively little of this production goes directly to consumers; rather, it is purchased in the form of completed houses, furniture, packaging materials, etc. Some products, like pallet lumber and wood chemicals, are used only in industry. Future demand for these wood products translates into a demand for trees on the stump. Since housing is such an important component of timber demand—an estimated 22 percent by the year 2000[22]—the size of the population and, particularly, the number of new households formed will have a major influence on demand. The level of overall economic activity will also be significant, especially in determining the demand for paper (44 percent of demand) and the level of nonresidential construction (26 percent).

Second, substitutes exist for wood in many of its uses. Steel, concrete, and aluminum often can be used in place of lumber for construction. Plastics can replace paper in many uses and often replace wood in furniture construction. The extent of this substitution depends partly on technology, but more on how much these substitutes cost.

Unlike wood, most of its substitutes are produced from nonrenewable commodities. Their prices may rise as reserves are exhausted, or may fall as new sources of supply are discovered. The major wood substitutes generally are considered more energy intensive than wood, since large quantities of coal are required to produce steel, electricity for aluminum, petroleum for plastics, natural gas for cement, and so on. According to one study, the energy consumed to process a ton of lumber, rolled steel, and rolled aluminum stands in the ratio of 1 to 8.4 to 45.[23] Expressed in terms of energy per dollar's worth of each of these commodities, the advantage of wood is not so obvious.[24] If energy prices continue to climb, prices of wood substitutes will rise as well, although the behavior of their prices *relative* to wood is not certain.

Whether substitution is desirable depends on environmental as well as economic costs. For example, producing a ton of steel involves the removal from the ground (and ultimate disposal) of about nine tons of raw materials and generates substantial amounts of air and water pollution. Other substitute materials present similar problems. As yet, many of these environmental externalities are not reflected in prices of these materials, although the trend of environmental regulation indicates that this may not be true in the future.

Third, demand for wood depends upon its price. Higher prices encourage wood users to cut their consumption or to switch to substitutes. There is undoubtedly a price, in fact, at which timber demand in the future would be no higher than it is today. Simultaneously, higher wood prices increase wood supplies, as less wood is left in the forest after harvest, mill wastes are recycled, less accessible sites are cut, and new investments are made in tree planting and timber-stand improvement.

Since the beginning of the nineteenth century, the price of lumber (in constant dollars) has risen steadily, with the rate of increase averaging about 1.7 percent yearly.[25] The rise reflects the steady depletion of the most accessible timber stocks, a process still underway in the virgin forests of the Pacific Northwest. If the historical rate of price increase were to be maintained, "real" timber prices would approximately double by the year 2020. There is evidence, however, that public policy may not be willing to accept large future price rises, particularly when they occur as spasmodic price run-ups of great size. Twice in the last decade—in 1968-70 and 1972-73—timber price explosions have been followed by pressure from lumber producers and from the home-building industry to increase the amount of timber the Forest Service allows to be cut on the national forests.[26] As the middle-class dream of owner-

ship of a single-family home seems more and more beyond the reach of most Americans, continued political resistance to anything that tends to raise housing prices is likely—including increases in lumber prices.

Table 2 compares projected national demands for various types of timber with present levels. For future dates, two levels of demand are given. The first assumes that relative timber prices stay at their 1970 level; the second (in parentheses) assumes a substantial rate of price increases. In both cases, the outlook is for significant, but not dramatic, increases in demand.

One immediately apparent trend is that demand for pulpwood will increase faster than that for sawlogs—a fortunate situation, for sawlogs have been the scarcest type of timber in recent years and the one subject to the greatest price pressure. Moreover, pulpwood supplies are by their nature easier to increase than are sawtimber supplies, for trees can be grown to harvestable size in as little as 20 years, compared with the 30 to 100 years that it takes to grow a tree of sawtimber size.

TABLE 2

CURRENT AND PROJECTED U.S. CONSUMPTION OF TIMBER
(1970 = 100)

SOFTWOOD	1952	1970	1980	1990	2000
sawlogs and veneer logs	88	100	132(115)*	141(112)	147(105)
pulpwood	71	100	121(121)	150(153)	185(185)
misc. prod. & fuelwood	267	100	100(100)	100(67)	100(67)
HARDWOOD					
sawlogs and veneer logs	93	100	143(121)	157(121)	179(129)
pulpwood	30	100	170(160)	260(220)	350(310)
misc. prod. & fuelwood	317	100	100(100)	100(100)	100(83)
ALL SPECIES					
(U.S.F.S.)	94	100	128(117)	148(128)	171(140)
ALL SPECIES**					
(R.F.F.)	—	100	121	135	160

SOURCE: U.S. Forest Service, *The Nation's Renewable Resources: An Assessment* (Washington: U.S. Forest Service, 1975), p. 233. The projections are based on moderate growth in population and income and make allowance for expected changes in the prices of substitutes.

*The first figure is demand at 1970 relative prices. The figure in parentheses is demand assuming relative prices of forest products rise as follows: lumber 1.5 percent per year; plywood, miscellaneous products, and fuelwood 1.0 percent per year; paper and board 0.5 percent per year.

**Projections of Leonard L. Fischman, "Future Demand for U.S. Forest Resources," in Marion Clawson (ed.), *Forest Policy for the Future* (Washington: Resources for the Future, 1974), p. 64. Assumes a year 2000 population of 288,000,000 and an unspecified rise in price.

The relative increase in hardwood demand is also significant, both for sawlogs and pulpwood. This reflects the substitution of more plentiful hardwood lumber for increasingly expensive softwood lumber, as well as new technology that permits the pulping of many hardwood species.[27] The better grades of hardwood, already in great demand by the furniture industry, should continue to be sought after as that industry expands. Pallet manufacture, which uses the poorer hardwoods, should also be at high levels, although the adoption of reusable pallets may mean eventual market saturation.

What do these future demands mean for the eastern national forests and for the private lands in and around them? Almost certainly, they indicate that previously unsalable products of these forests will find ready markets. The greater use of hardwoods for both lumber and pulp will substantially improve the market for Appalachian and Lake States hardwoods. These are now quite plentiful, with growth far exceeding cut on both public and private lands. Demand is unlikely to exceed the supply of these hardwoods in the foreseeable future, with the significant exception of large trees of species used in furniture and cabinets, which already have risen considerably in price.

Geographically, the depletion of softwood timber stocks on industry lands in the West will shift the focus of cutting eastward, particularly to the deep South. One industry-sponsored study predicts that, by the year 2000, the South will provide the majority of the nation's wood supply, producing more than twice as much as the region does today.[28] More than half of that production, both hardwood and softwood, will be pulped. Already, major timber companies have purchased millions of acres of southern land and planted pines, some of them genetically selected "super-trees."

Prospective increases in demand and price imply that there will be more lumbering on the private lands in and around the national forests.[29] This will prompt more requests for road access to inholdings and the need to mark national forest boundaries to identify public and private timberlands. Erosion from logging and pollution from new paper mills are also likely.

Greater demand for timber (and higher timber prices) will create new investment opportunities, both within the eastern national forests and on the private lands. This could result in increased planting, more frequent thinning cuts, and the conversion of existing stands from one species to another—in short, more intensive management. In the South, predicts the Southern Forest Resource Council:

> Through intensification of forest management, most of the oak-pine forest type and much of the oak-hickory on average or better sites is likely to be converted to pine. Pine and oak-pine types, now about 104 million acres, are expected to increase to 118 million acres. The approximately 95 million acres of hardwood types in 1963 may be reduced to about 70 million acres in the year 2000.[30]

National forest investment does not necessarily respond to market stimuli, but even there the creation of possibilities for increased revenues is likely to invite pressures to invest in more intensive timber-management practices.

The wholesale conversion of private lands from hardwood to pine may, paradoxically, increase the incentives for producing hardwoods on the national forests. According to one Forest Service document, such conversion on private lands in the Ozark region could result in an eightfold increase in the demand for national forest hardwoods between now and 1990.[31] This may be compounded by the recent trend toward clearing and draining of bottomland hardwood sites in North Carolina and the Mississippi Valley for use in cotton, soybean, and rice production. Says a forest industry economist:

> These are some of the best hardwood sites in the United States. . . . If we lose good quality growing sites in bottomland hardwood land, we will need national forest timber because of its high quality.[32]

Finally, the almost inevitable price rises that will accompany future demands will probably cause a new round of proposals for greater timber production from the national forests. Such a translation of economic pressure into political pressure is predicted by a timber industry trade journal:

> Over the long run . . . there is much concern over the danger of a lumber and plywood shortage once housing production takes off. . . . With production held back by failure to develop national forest harvests to their potential and a lack of encouragement for owners of smaller woodlots, another crisis similar to—or worse than—that of a few short years ago, with homebuilders demanding government control of lumber prices, may well occur again.[33]

Moderation of future lumber price rises by increased cutting on the national forests was the thrust of the President's Panel on Timber and the Environment (1973). Some increase in national forest production is a component of nearly all of the "program alternatives" proposed by the Forest Service's Resources Planning Act Plan (1975). According to the latter:

> Timber utilization opportunities could increase national forest system timber sales from the current level of 12 billion cubic feet to as much as 28 billion cubic feet.[34]

Large and rapid increases in production would have to come from a faster liquidation of the "old growth" forests of the western states, which currently contain about one-half of the nation's entire stock of softwood sawtimber. As a complement to this effort, and to provide for longer-term wood supplies, however, many advocate what the timber panel calls "intensified management on the most productive forest sites and types."[35] Neither the timber panel nor the RPA program gives any detail on how such a management effort could be allocated geographically. But if a "productive site" is defined as one capable of producing 85 or more cubic feet of wood per acre, about 3,800,000 acres of eastern national forest land (16 percent) would meet the standard.[36] Nearly three-quarters of this land is in the South.[37]

The composition of demand also has implications for cutting methods. If there are high levels of demand for large, furniture-quality hardwoods, selective cutting of the larger trees may become more attractive to loggers. This would also be true if prices of lumber increased. On the other hand, the increasing demand for pulpwood may mean ever greater removal from the forest of small-sized trees which previously either were not cut or were left behind as waste. This would make clearcutting even more profitable than it now is.

Many foresters claim that clearcutting—although not necessarily in large blocks—is the only way to grow shade-intolerant species such as the southern pines, aspen, jack-pine, yellow poplar, and cherry. The relative demand for these, as opposed to such tolerants as hickory, white oak, and sugar maple, helps determine whether clearcutting is financially desirable.

RECREATION

Recreational use of the eastern national forests ranges from such vigorous pursuits as skiing and whitewater canoeing to relaxed activities such as picnicking and pleasure driving. Activities like snowmobiling or trailer camping require the participant to bring a great deal of equipment into the forest; others require none at all. Accommodating an activity may mean that the Forest Service actively provides the opportunity—a campsite with water and toilet, a road or trail, a managed habitat for game or fish. It may, on the other hand, mean that something is *not* done with the forest—a streamside is not clearcut, a road is not pushed into a wilderness area. Some kinds of recreational pursuits require high-quality scenery. Others need only a tree-covered area. Still others do not really need a forest environment at all.

In 1974, visitors spent some 44,000,000 days engaging in some form of recreation in eastern national forests (Table 3), for an average of slightly less than two visitor days per acre per year.[38] Activities were, of course, unevenly distributed among the forests and among individual sites within the forests. Yearly visitor days were as high as 6 per acre on the Wayne-Hoosier (Ohio-Indiana) Forests and 3.6 on the White Mountain (New Hampshire), while Louisiana's Kisatchie and Wisconsin's Chequamegon registered just 1 per acre or less. There is probably even greater variability in intensity of use for particular pieces of land within an individual forest. Even in the heavily used White Mountain, for example, many areas are little visited. Use also varies considerably with the seasons; campsites and trails overcrowded on summer weekends may be virtually deserted on weekdays or in other seasons.

TABLE 3

OUTDOOR RECREATION IN EASTERN NATIONAL FORESTS 1974

Activity	Visitor Days	Eastern Forests as Percent of Total National Forest Use
Mechanized Recreation Travel	11,000,000	25
Camping	9,500,000	19
Hunting	5,900,000	41
Fishing	5,100,000	31
Boating	2,100,000	37
Hiking and Mountain Climbing	1,900,000	23
Viewing Scenes, Sports, Environment	1,600,000	26
Picnicking	1,500,000	22
Swimming and Scuba Diving	1,300,000	33
Interpretation	703,000	21
Winter Sports	665,000	8
Recreation Residence Use	594,000	9
Gathering Forest Products	495,000	23
Horseback Riding	450,000	16
Organization Camp Use	430,000	10
Resort Use	379,000	10
Waterskiing and Other Water Sports	303,000	26
Nature Study	212,000	21
Games and Team Sports	120,000	16
Total Use	44,000,000	23
Total Land Area in Eastern National Forests (%)	24,000,000 acres	(13)

SOURCE: U.S. Forest Service, *Report of the Chief*, 1974.

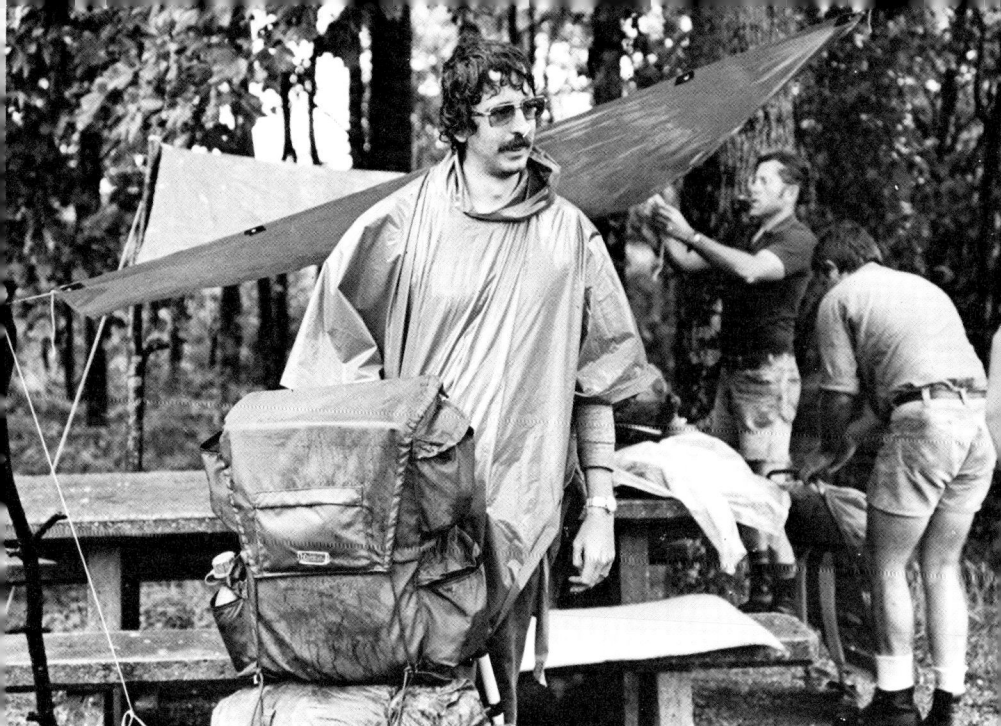

We like the Appalachian Trail. One of the great things is that you find people of all ages out here. Where else would you find a person 51 years old and someone 14 or 15 doing the same thing?

You don't run into a lot of people who are tearing things up and yelling. You have to have the trail on your mind instead of the radio and the ball game, or liquor, or the things you can do somewhere else. In the 15 years I've been hiking on the trail, I haven't run into more than about two people who were rowdy. Remember? Three boys asked us yesterday if we minded if they turned on their portable CB radio. "Can we switch on the radio, or did you leave home or get away from them?" they asked us. We told them we didn't want to listen to it.

We come up here just to get away. You sit in an office all week, put up with the hassle—everybody's troubles. Up here you get away from people, cars, smog.

CAMPERS AT WOODY GAP in the Chattahoochee National Forest, Georgia

---

Recreation use throughout the National Forest System has been rising at a rapid rate for many years, and has increased more than tenfold in the postwar years. The most recent years have seen some slackening in the rate of growth, both in absolute and percentage terms, but it remains high. Moreover, the growth rate is higher on the eastern national forests than elsewhere in the system. Between 1967 and 1974, visitor days rose 43 percent in the forests of the East and South. This was far greater than the rate of population growth (6.6 percent), of recreation growth on the national forests nationwide (29 percent), or of visits to the national parks (23 percent).[39]

Forest Service statistics (Table 3) indicate that eastern national forest visitors spend the greatest part of their time driving through the forest, whether on a surfaced highway, a graveled road, or a jeep track. Some of this is sightseeing for its own sake; much is incidental to other forest activities. For many visitors, then, the view of the forest extends no farther than the view from the road. Next in importance among forest recreation activities are camping, hunting, and fishing. Water sports also enjoy considerable popularity, both on natural lakes and on the reservoirs of such forests as the Allegheny, Chattahoochee, and Sabine.

The Forest Service classifies recreation uses as "dispersed" and "developed." The former take place at low density and can use much or all of the forest. These uses include hunting and fishing, hiking, primitive camping, and wildlife observation. Developed uses involve high densities and the necessary provision of certain facilities (principally sanitation, water, and refuse collection). It has long been the Forest Service's policy to charge fees only for the use of special facilities, such as developed campgrounds and boat ramps.[40] In 1974, these recreational fees on the eastern forests raised only $1,200,000.[41]

In the past, the Forest Service tried to satisfy the public demand for both kinds of recreation, even when it meant building and managing facilities far from rustic. More recently, policy seems to have shifted in favor of dispersed recreation. "Area guides" for the eastern forests include such phrases as "emphasize dispersed recreation as the primary recreational experience"[42] and "the major emphasis throughout the Coastal Plain area will be dispersed recreation."[43] Dispersed recreation is also favored in the Forest Service's recommended recreation program under the Renewable Resources Planning Act. This suggests a recognition that developed recreation is expensive and difficult to manage and that elaborate facilities perhaps are more appropriately supplied by private business on inholdings or just outside the forest boundaries.

But the Forest Service's desire to accommodate the whole spectrum of public recreation demands seems to be dying hard. In 1975, a high Forest Service official told a Congressional committee:

> In the future, new Forest Service-operated recreation sites will be provided in areas which represent a broad spectrum of recreation experience. This spectrum is reflected in five levels of recreation facility de-

velopment and environmental modification—from primitive to modern. Emphasis in facility development will favor larger sites located closer to population concentrations and often associated with water-oriented environments. Larger sites lend themselves to more efficient sewage treatment, more adequate public protection, and improved control of vandalism.[44]

The most highly developed recreational facilities on eastern national forests include ski developments built by private business under Forest Service permit. Several popular ski areas in Vermont and New Hampshire, including Mt. Snow, Wildcat, and Sugarbush, are within national forest boundaries. Typically the buildings and commercial facilities occupy a private inholding, while the ski slope itself is on a mountainside owned by the Forest Service.[45] It appears, however, that public lands are not nearly so important to the ski industry in the East as in the West, where national forests account for 80 percent of the major ski areas.[46]

In the past, the Forest Service issued permits for private recreational residences on public land, charging a yearly fee for that privilege. Current policy is to issue no new permits, gradually phasing out this use on some forests but not on others. While not a major element in forest recreation, such homesites are not uncommon in forests in Florida, Minnesota, and Tennessee.

At the other end of the recreation spectrum, there has been considerable debate over the creation and future management of wilderness areas in the eastern forests. The original Wilderness Act, passed in 1964, limits the lands designated to those "generally appearing to have been affected primarily by the forces of nature, with the imprint of man's work substantially unnoticeable." Almost immediately, wilderness advocates and the Forest Service began to argue over what Congress meant by these words. This "purity" debate had special relevance to eastern forest lands.

Heavily logged in the past, laced with roads, perhaps once farmed or grazed, most of the land in the eastern forests at some time in the past had been significantly affected by human activity. At issue was whether the passage of time and the healing powers of nature had so obliterated man's traces on the land that it regained its "wilderness" quality. The Forest Service proposed a strict definition of wilderness, in part fearful that anything less would open almost any land in its possession to possible designation. The Service took the position that it was advocating a high standard for wilderness. According to Chief John McGuire, "Should [restored eastern forest lands] be considered as primeval, vast areas within the various federal land systems would also qualify. The uniqueness of the present wilderness system would disappear."[47] Wilderness advocates, on the other hand, asserted that "the issue is not whether the area ever was disturbed, but only the more practical judgment whether the evidence of past disturbance is not substantially unnoticeable."[48]

From a personal standpoint, I would like to see as much wilderness as possible. But, of course, I realize that people have to make a living through timber and other forest resources. So there have to be some compromises.
DANIEL STILLWELL, Professor of Geography, Appalachian State University Boone, North Carolina

---

It seemed that without some resolution of the "purity" issue, little eastern land could be designated under the Wilderness Act. In fact, between 1964 and 1974, only four wilderness areas were designated under the act on eastern national forests: relatively small tracts at Linville Gorge and Shining Rock, North Carolina; Great Gulf, New Hampshire; and the million-acre Boundary Waters Canoe Area in Minnesota.[49]

Late in 1974, Congress passed the Eastern Wilderness Areas Act, designating 16 eastern national forest areas, containing some 207,000 acres of land. They ranged in size from the 34,000-acre Cohutta Wilderness in Georgia and Tennessee to the 2,570-acre Gee Creek Wilderness in Tennessee. An additional 17 tracts (125,000 acres) were named as study areas, with the Forest Service required to make recommendations by 1980 as to their potential for inclusion in the system.[50] While they are being studied, the Forest Service is required to manage them "so as to maintain their presently existing wilderness character." The new law made no mention of the criteria used for designation, although wilderness advocates took the inclusion of these places in the wilderness system as evidence that Congress agreed with them that regenerated areas could still contain wilderness values. In 1976, Congress designated a 12,000-acre national forest wilderness in Missouri, as well as four new Missouri study areas totalling 28,000 acres.

To date, wilderness designation has received more public attention than wilderness management. Although wilderness areas are protected from timber cutting and road building, there are unresolved issues involving the closing of existing roads and removal of huts and toilets, the granting of mineral leases, and the extent to which an area will be protected from fire and insects. Perhaps the most pressing management issue concerns the Boundary Waters Area, created under a special section of the 1964 Wilderness Act, which requires that its primitive character be maintained "without unnecessary restrictions on other uses." There has recently been controversy over a Forest Service plan for the Boundary Waters Area that would allow continuation of timbering and snowmobile use within part of the designated area.[51]

No fee is charged for wilderness use, but permits are required for 10 of the 20 eastern wilderness areas. There is currently no limitation on the number of permits issued, except at Linville Gorge, where only 60 overnight visitors are accommodated at one time, and at the Boundary Waters Canoe Area.[52] Wilderness proponents generally have not objected to rationing by permit to protect the resource and to enhance the quality of a wilderness visit.

Perhaps the most controversial recreational uses of the eastern national forests involve snowmobiles and other motorized vehicles which either go cross-country or use primitive roads and trails. These have been among the fastest-growing outdoor sports in recent years. Such vehicles are associated with soil compaction, damage to vegetation, interference with wildlife, and disturbance to less noisy users of the forest.[53] In February 1972, a Presidential executive order called for regulation of off-road vehicles on the public lands. The Forest Service's off-road vehicle regulations, which were issued in 1973, have evoked heated protests, both from environmentalists, who consider them too weak, and from vehicle users, who see them as unduly restrictive.

The motel owners have tried to operate year-round by emphasizing skiing. It didn't work. Then we got together to see if we couldn't organize something around snowmobiling. We talked to the Forest Service and put together a system of snowmobile trails. It got to be very big; use of the national forest increased. This has had a big economic impact on Twin Mountain and some of the surrounding towns.

In 1969, we would have one motel, a restaurant, and two grocery stores open on a winter weekend—and that was all. Now we have about 20 motels, four service stations, three grocery stores, and two campgrounds open on a year-round basis. Motel owners have spent a lot of money winterizing their facilities. Now a lot of people who used to leave town in the winter are staying here.

When you talk about economic value, snowmobiling is unsurpassed. It's

an expensive sport and spreads a lot of money around. The hikers bring their own food and stay in an Appalachian Mountain Club Hut. Snowmobilers stay in local motels, eat at the restaurants, patronize the gas stations.

Everybody who wants to use the national forest should be able to—under controlled conditions, of course. Everybody likes to move over the ground in his own way. Some like to walk, but others like speed. Some are older people, and can't get around in the woods any other way. Anybody with any ability at all can ride a snowmobile.

We have run into small conflicts, but we can iron them out. Hikers and ski tourists want quiet for their activities, and as far as I'm concerned, they are justified in that. In the few areas where there is real conflict, we're working on alternate routes so that hikers, ski-tourists, and snowmobilers can have their privacy and use the forest the way they want to. That's what the forest is for.

HAROLD GARNEAU, Twin Mountain Snowmobile Club
Twin Mountain, New Hampshire

Outdoor recreation activities tend to be pursued either by young people or by family groups. Now that the postwar "baby-boom" generation is reaching adulthood, there is no longer such a large built-in demographic impetus to increases in demand. Future recreation demands will depend on whether the early exposure of the baby-boom generation to outdoor activities will make them more likely than their parents to continue to engage in outdoor pursuits as they get older. It also will depend on the number of children that this generation has.

Table 4 presents Forest Service projections of national recreation demand in the coming decades. These indicate that recreation demand will increase significantly, but not explosively.

Off-road vehicle users can't understand that the experience of a ski tour or a snow-shoe trip is damaged by snowmobile activity. They don't understand there are two kinds of outdoor recreation. One is passive and the other active. It's like the water skier and the fisherman. The water skier goes by blissfully and waves to the fisherman, but the fisherman is angry because his fishing's been destroyed.

So the snowmobilers can't understand why some areas are closed, screwing up their trail systems so that they can't easily get to one town, or get to their grandiose north-south New Hampshire trail, because certain small areas in the heart of the national forest are a barrier to through traffic. I've heard snowmobilers get up time after time and say, "My country 'tis of thee. Freedom for everybody," and "Why can't I drive my snowmobile into that notch?" And they can't understand how people really need to be away from machinery and certain areas have to be closed for more passive kinds of recreation.

I've explained that there's a real need on the part of people who travel from afar to use this forest for some solitude and respite from urban-type influences—noisy machinery, and so forth. This is a national forest, and it belongs to all the people of this country. I've gotten that point across well one or two times in public hearings, but other times I've just been a threat, like when I've just listed areas that we wanted to be closed—like Wild River, Carrigan Notch, Shoal Pond, Mountain Pond. I think in most cases the snowmobilers won out because their argument was that there're only certain places where they can go through the White Mountains, because they're limited to the kinds of terrain they can use. So they say, "We just want to use three main through-routes." But in some cases, those turn out to be right through the middle of a valley that's otherwise wilderness or real high-quality ski touring, snowshoeing, or backpacking. They eliminate a lot more land area than they realize by just having a single through-route, just because of the acoustical nature of snowmobiles.

ROBERT PROUDMAN, Trails Supervisor, Appalachian Mountain Club Pinkham Notch Camp, Gorham, New Hampshire

TABLE 4

PROJECTED NATIONAL DEMANDS FOR OUTDOOR RECREATION BY THE YEAR 2000

Substantial Increase (45% or more)	wilderness use, sailing, freshwater fishing, nature photography, big-game hunting, water skiing
Moderate Increase (20-45%)	off-road driving, remote camping, developed-site camping, hiking, swimming, picnicking, horseback riding, bicycling, nature walks, birdwatching, sightseeing
Small Increase (20% or less)	small-game hunting, waterfowl hunting

SOURCE: U.S. Forest Service, *Nation's Renewable Resources: An Assessment*, compiled from "medium" projections.

If present trends continue, the eastern national forests are less likely to be confronted with another huge expansion in the *quantity* of recreation demanded than with significant changes in *types* of use. Future recreation demand will tend to emphasize activities requiring outdoor skills and particular kinds of forest environment rather than merely a wooded plot of ground. If the rate of population growth were to increase significantly, of course, there would be a corresponding increase in the more common types of recreational pursuits—camping, swimming, hiking. A second baby boom would produce, after a few years, a new explosion in outdoor recreation demand.

Of particular significance for management decisions is the projected increase in demand for wilderness land. The Forest Service quotes research which suggests that between one-fourth and one-half of all wilderness users might be satisfied with a roadless recreation area rather than a true wilderness.[54] Whether or not this is true, advocates of wilderness in the East are far from satisfied with the amount of land now devoted to this use. One review of potential areas, prepared by the Sierra Club, lists (in addition to the designated areas and study areas in the 1974 Eastern Wilderness Act) some 42 tracts of eastern national forest land, totaling 721,000 acres.[55] To put this in perspective, were all of this land added to that already designated or under study, wilderness would account for 8.4 percent of the Service's 24,000,000 acres of eastern forests.

The growth of some specialized recreational activities will place severe demands on forest resources. Already the whitewater craze has overloaded some of the most popular rivers and presented problems of how to protect participants who lack the necessary skills. Cliff faces have been scarred by the pitons of rock climbers and, in some places, excessive wilderness use is destroying wilderness values. Even such activities as wildlife study, rock hunting, and collecting of plants and other forest products may have to be managed to prevent damage to the resource itself.

In the future, there probably will be as many conflicts between types of recreational uses as there now are between recreation and the other multiple forest uses. Sailors, fishermen, and water skiers will demand more water impoundments, while wild-river enthusiasts will urge that streams be kept free flowing. Snowmobilers will conflict with cross-country skiers. Those who enjoy scenic drives will want greater road access, while others, seeking solitude, will seek the closure of existing roads. Those who come to observe wildlife will vent their distaste of those who come to hunt it.

In the future, I see increasing demands upon the resources of the forest, and not only for timber. We already see all types of recreational demands: hunting, fishing, simply traveling through the forest. Then there are the picnickers and sightseers who never go into the forest itself but like to look at it. Go over to the Swift River on a pleasant day and see the hundreds of people just sitting there looking at the rocks and the water. We'll see an increasing demand for roadside conveniences. About one and a quarter million people drove over the Kancamagus Highway last year. Think of a two-lane secondary highway accommodating such a volume of traffic.

SHERMAN ADAMS, Owner, Loon Mountain Ski Resort
Former Governor, State of New Hampshire
Lincoln, New Hampshire

An increased demand for certain types of developed recreation is one of the consequences of an aging, more affluent population. Resorts, golf courses, second homes, and small boat marinas are only indirectly related to less intensive forms of outdoor recreation, but they also demand pleasant natural settings. These facilities, less likely to be placed on national forest land than on privately owned inholdings and adjacent lands, may create environmental problems, such as water pollution and fire dangers, that can spill over onto federally owned land. More importantly, those who build and use them will make increased demands on the public lands—for road and utility rights-of-way, for recreational use, and for modifications in management practices. Many forest managers would agree with one forest supervisor who notes "a tendency for these [adjoining] landowners to take the attitude that the surrounding public land is their own private playground and should be managed to enhance their property."[56]

The future supply of recreational development is difficult to forecast, particularly because of uncertainty about energy prices, interest rates, and state and local land-use controls. Nevertheless, unless prices become extremely high, the demand will continue. Currently, only about five percent of U.S. households own a second home, a very low degree of market penetration.[57] As oceanfront and lakeside properties become increasingly scarce in the eastern states, more developers probably will turn to private lands in mountain areas, many of them in or near national forests.

WILDLIFE

The eastern national forests support an impressive quantity and diversity of wildlife. The northern forests alone provide habitat for members of 9 endangered or threatened species and an additional 28 which "need management to prevent them from becoming endangered."[58] These include the eastern timber wolf, the northern bald eagle, the great blue heron, and such humbler creatures as the Indiana bat and the Ozark cavefish. The Forest Service spends about $3 million a year on wildlife management in the eastern forests, principally on habitat improvement and coordination of habitat management with timbering and other uses.

A small amount is budgeted for efforts to conserve endangered species. Perhaps the most notable example is in the 4,000-acre Kirtland's Warbler Management Area on the Huron National Forest in Michigan, where the Forest Service burns several hundred acres of land yearly, for the warbler nests only at the base of young jack pine—a fire species. Most of the wildlife budget supports management for the more common, and huntable, species, such as deer and turkey. At the time of the Weeks Act purchases, even these common species had been almost eliminated on some of the eastern national forests. It is said, for example, that not a single deer remained on the George Washington National Forest in Virginia when it was purchased in the 1920's. (This may be an exaggeration, but there is no doubt that deer herds had been seriously depleted.)

Gradual reforestation, cuts designed for wildlife purposes, physical reintroduction, and protection by state game laws have combined to restore sizable populations of deer to the George Washington. A similar success story is the relatively recent reestablishment of the wild turkey on national forests throughout the East.

Hunting and fishing are permitted on almost all lands within the eastern national forests, subject to state laws. Each year hunters kill about 75,000 deer on these forests, with most hunting taking place in the forests of Virginia, Michigan, Pennsylvania, and Minnesota.[59] Because a large amount of the private forest land in the East is closed to hunting, the national forests probably provide a greater proportion of the large game kill than their acreage would indicate.[60] In some areas, the Forest Service has been under pressure to increase its game supply. But its response has been limited, since overabundant deer populations eat young hardwood seedlings and make it difficult to regenerate some timbered sites. Encouraging the deer population to increase beyond the land's grazing capacity also, of course, makes survival difficult for the animal.

Deer is a crop in Pennsylvania—let's face it. The more deer you harvest, the more deer you have. This isn't what some conservationists would like you to believe, but hunters really save the deer from themselves. Deer harvesting is just like harvesting cattle. You can't have a hundred head of cattle in an area which can support only five cattle. By the same token, you can't have 2,000 deer in an area which can support only 100. When deer are killed, it makes room for more deer next year, because the deer have more food and the does are in better shape, and they produce two fawns instead of one.

JACK SKINNER, Service Center Foreman, Bell Telephone Company
Warren, Pennsylvania

Fish are stocked in national forest streams by state agencies. There is no firm policy emphasizing the stocking of native species, although exotic species are discouraged. Stocking is permitted in wilderness areas, provided it does not involve the use of vehicles.

The analysis of future recreation demands on the forests indicated that some of the consumptive uses of wildlife (big-game hunting, freshwater fishing) and all of the nonconsumptive uses (bird-watching, nature photography, nature walks) are expected to increase in the future. As private forest lands are cleared for agriculture or modified for urban or recreational use, federal lands will

A wild turkey and its young scurry across a forest road in Virginia's Jefferson National Forest.

become more important as wildlife habitat, particularly for species that can survive only when isolated from man and his noisy artifacts. The projected conversion of southern lands from hardwoods to pines will have adverse effects on such species as squirrel and turkey and will give greater importance to the hardwood stands remaining on the national forests.

WATERSHED

The stated purpose of the Weeks Act was protection of "the watersheds of navigable streams." In the early 1900's, the act's proponents well understood the influence of uncontrolled cutting and burning on soil erosion and downstream flooding. Only recently they had seen evidence of it in frequent floods on the Connecticut River downstream from the White Mountains, and in the disastrous 1907 flood on the Allegheny and Monongahela Rivers.

Since that time, reforestation and the suppression of forest fires have largely eliminated these lands' potential for generating serious floods and siltation. In recent years the Forest Service's watershed program has emphasized managing other forest activities so that they will not unduly impair water quality. This includes restricting timber cutting along stream banks and on steep slopes, seeding or otherwise reforesting newly cut areas, and setting erosion standards for forest roads. There have been notable failures of these policies in the past, such as the erosion from ill-constructed timber roads and large-scale clearcuts on the Monongahela National Forest,[61] but in general the quality of water in the eastern national forests is quite high.

Two kinds of exceptions mar this success record. Where mining is practiced on public land or inholdings, mine-acid drainage has been a problem on some forests. This has particularly affected the Monongahela, the George Washington, and the Wayne. Sewage discharges from towns or other developed private lands within forest boundaries also present problems of both health and aesthetics.

Beyond taking what are essentially defensive measures, the Forest Service does little actively to manage the *quantity* of water flowing from its eastern forest lands. Although its research shows that manipulation of vegetation type can increase water flow from forest land, funds are limited for implementing this work, and little demand now exists for it in the water-rich East and South.

Land acquisitions in recent years have emphasized forest values other than water production, as a review of recent land purchases under Weeks Act authorization reveals. Section 6 of the act requires that the Secretary of Agriculture approve lands purchased for timber production, while lands purchased that "will promote or protect the navigation of streams" must also be approved by the Director of the Geological Survey. In the past few years, the decline is notable in the number of land purchases necessitating the approval of the Director of the Geological Survey.[62] Purchases of recreation lands under the Land and Water Conservation Fund frequently have included stream banks and lakeshores, increasing the Forest Service's watershed ownership, but for reasons very different than those contemplated by the Weeks Act.

Eastern national forest lands provide some 170 municipal watersheds, with the greatest number in New Hampshire and Virginia. National forests in the east do not furnish the water supplies for large cities as do many western national forests. Rather, these eastern forests tend to provide the headwaters for a river system, and not the major part of its flow. Among the rivers that have headwaters or tributaries on eastern national forest lands are the Connecticut, Ohio, Arkansas, Mississippi, Tennessee, Potomac, James, Chattahoochee, New, and Ouachita.

While the nation as a whole is expected to face future water-supply problems, few of these will affect the East or the eastern national forests.[63] But these forests do have a future role in improving water quality. Many of the major stream systems of the East—including the Ohio, Arkansas, and Potomac—are badly polluted. Clean water from tributaries originating on national forest land helps dilute the wastes generated by industry and population centers. Moreover, at present levels of timber cutting, sediment loads from national forest lands are much lower than those generated by other land, especially farmland. Anything that would change either the purity of eastern national forest runoff (e.g., mine-acid drainage) or the amount of sediment it carries (e.g., erosion from forest roads) will tend to reduce the forests' current contribution to the quality of major streams.

National forest lands may come under pressure for water-impoundment projects in the future, although perhaps less so than in the past. The trend toward requiring greater economic justification for projects built by the Federal Government probably will reduce their total number, while greater attention to environmental impacts probably will limit their location on national forest land.

RANGE AND GRAZING

About 15 percent of eastern national forest lands are grazed by livestock, principally by beef cattle owned by farmers and small ranchers. To a much lesser extent, grazing is also used for aesthetic purposes and to improve habitat for game birds. In the Mount Rogers area of the Jefferson National Forest, for example, a small amount of sheep grazing is encouraged to maintain scenic clearings kept open in earlier times by uncontrolled fires.

The eastern national forests supply about a million "animal-unit months" of use, far less than half of one percent of the rangeland grazing in the country. Nearly all of the grazing takes place on the pine forests of the South, mainly in Arkansas, Mississippi, and Louisiana. As late as 1965, most of this use was by trespass. Trying to control such use, the Forest Service imposed nominal fees for grazing permits. Now that it has successfully reduced trespass grazing, it is increasing these fees toward market levels.

Despite the fact that the eastern national forests are rather prolific producers of forage material (averaging 817 pounds per acre), they have not been nearly so heavily grazed as their western counterparts. A 1972 Forest Service study showed that eastern national forest lands could support perhaps four times the level of current grazing.[64] Use has remained below its potential because of obstacles that include low levels of funding for management, fencing, and water supply; the seasonality of natural forage; and, until quite recently, a lack of effective demand.

Demand has been low because of the ready availability of forage on private pasturelands and the lack of growth of the cattle population.[65] Now, as the number of cattle rises and idle lands are put back into crop production, the forest range looks increasingly attractive to livestock producers. This is especially true in the South, where the cattle population was formerly limited by disease and parasites. The South now, says a Forest Service range specialist, "has the fastest growth of beef cattle in the nation."[66] In coming years he says, most of the nation's 30,000,000 additional cattle could be raised in the South. He considers the forest understory there "a whole new untapped resource," and also sees major long-term potential for grazing in some of the northern forests, particularly on cutover land on Lake States forests such as the Hiawatha and Superior. Indeed, figures show that some hardwood ecosystems, such as aspen-birch and maple-beech-birch, are the largest potential forage producers on the eastern national forests.[67]

The Forest Service is starting to take a greater interest in the range use of the forest understory, and asserts that relatively large increases in grazing can be achieved with little interference with other forest uses.[68] Other observers are not so sure; they cite the value of forage and cover to deer, turkey, and other wildlife, and the damage to young trees that grazing can cause, particularly in the hardwood ecosystems.[69] Other possible detrimental effects of increasing grazing on the eastern national forests could include aesthetic impacts of fencing and structures, damage to stream banks and beds by animals, water pollution, sedimentation, and problems from herbicides and fertilizers.

MINING AND MINERAL EXTRACTION

Mining is an important use on a few of the eastern national forests; most are currently untouched by it.[70] In 1975, 56 percent of the lead and 12 percent of the zinc mined in the United States came from deep mines on the national forests in Missouri. Substantial amounts of fluorospar, a mineral used in making industrial acid, are taken from the Shawnee National Forest in Illinois. Oil and gas are produced on the Allegheny National Forest in Pennsylvania and on the Homochitto and DeSoto National Forests in Mississippi. A commercial gravel mine is operated on the Mark Twain

National Forest in Missouri, and state and county authorities are allowed to take sand and gravel for construction from other forests. Coal is surface mined in Ohio's Wayne National Forest. But the existing level of mineral production is negligible compared to future possibilities.

The extraction of national forest mineral deposits is governed by an exceedingly complex set of laws. The Forest Service's ability to regulate mining and its environmental impacts depends in part on whether the mineral to be extracted is a "common variety" (sand, gravel), a "hardrock mineral" (uranium, lead), or a "leasing act mineral" (oil, coal, phosphate). It also depends on whether the particular piece of forest land was created out of the public domain and, in the case of the acquired eastern forests, whether mineral rights were reserved by the owner when the Forest Service bought the land or were already "outstanding" (in the hands of third parties) at the time the transaction was made. If the rights are reserved, the land is subject to whatever set of environmental regulations was in effect at the time the land was purchased by the Forest Service. There are four sets of such regulations, representing the state of environmental consciousness as of 1911, 1937, 1947, and 1963.

In many of the highly mineralized areas, the government must contend not only with the spillover effects of mining activities on inholdings and adjacent lands, but with the fact that it owns the mineral rights under only two-thirds of its own land. Ownership of the other one-third of all the government's Weeks Act holdings lies in private hands.[71] The proportion in private ownership is higher in those areas where minerals are known to be present. For instance, within the Monongahela National Forest, in West Virginia's coal country, the government owns 50 percent of the land, but only 28 percent of the subsurface. Similar situations occur on the Daniel Boone, the Jefferson, and the Wayne. In the Cranberry Backcountry, a wilderness study area within the Monongahela, the situation is even more dramatic, with private parties holding mineral rights on 95 percent of the land. Under it are coal deposits worth hundreds of millions of dollars. And on the Allegheny National Forest, in the heart of Pennsylvania's oil region, the government owns 68 percent of the surface, but less than 2 percent of the mineral rights.

Increases in oil prices have made it profitable t old well fields on the Allegheny National Forest in Penns

As the surface owner, the government has few rights. A deed covering most of the Cranberry area allows the owner of mineral rights to:

> enter upon and under said lands and to mine, excavate and remove all of said coal and other minerals . . . and also the right to . . . make and construct all necessary structures, railroads, roads, ways, excavations, air shafts, drains, and openings necessary or convenient for mining and removal of the said coal . . . without being liable for any injury or damage done thereby to the overlying surface or to anything therein or thereon, or to any watercourse therein or thereon.[72]

Owners of "outstanding" mineral rights can remove minerals, even by strip mining, with the land protected only by state strip-mining and water-pollution control laws. In many mining states, these laws are woefully inadequate or underenforced.

The Forest Service owns the surface rights on the Allegheny, but the subsurface rights are outstanding in third parties. Of the 500,000 acres that the national forest owns, we own only a little bit less than two percent of the subsurface. Therefore, we have no control over what the subsurface owners do to the surface to gain access to their holdings. In essence, the subsurface owner considers our ownership of the surface subservient to his interests.

We set standards, but the problem is that, whereas in agriculture—say corn or soybeans—crop rotations are only one year, in forest management we're talking about 80 to 120 years. And once they destroy the surface in a forest environment, it takes so long for it to come back that we're talking about somebody else's lifetime.

ROBERT L. FIELDS, Land Staff Officer, Allegheny National Forest
Warren, Pennsylvania

The Forest Service has considerably more power in cases in which it owns the mineral rights as well as the surface. It can impose conditions on how minerals may be extracted, or it may bar mining entirely. In a few cases, it must contend with the results of past commitments, some of them made in times of less concern about the environment. For example, in northern Florida, on the Osceola National Forest, four chemical and mining companies want to be granted "preference leases" to mine phosphate deposits which they have discovered. Some question exists as to whether the government has the power to revoke grants made under the less stringent regulations of previous years.

Mining clearly represents the greatest foreseeable threat to other uses of the eastern national forests. It is a land-intensive activity, one which could involve as many as 500,000 acres of forest land.[73] (By contrast, developed recreation sites now take up about 60,000 acres, even if their peripheral areas are included.) Mining can, and often does, result in great and even irreversible environmental damage, including acid drainage from underground coal mines, leachate pollution from heavy metal mines, contamination of subsurface water sources from oil drilling, and the literal remaking of the landscape. Mining is not a distant or theoretical threat, but a very current one. Considerable exploratory work now is underway, in addition to the continuing extraction of oil, coal, and other minerals. But past and current difficulties and conflicts involving mining are minor compared to those likely to occur as demand increases for minerals under federal lands.

High prices and strong demand for fossil fuels, metals, and phosphates (used in fertilizers) have motivated mineral companies to take a new look at the resources underlying the eastern United States. In some cases, this involves areas where the existence of minerals was long known or suspected, but the deposit was uneconomic to work at prevailing market prices. This is the situation on Pennsylvania's Allegheny National Forest, where the recent trebling of oil prices has made it profitable to redrill oil fields abandoned before the turn of the century. In other cases it involves exploring for new deposits, for even in the long-settled East, mineral geology has many gaps. As recently as 1955, for example, a whole new body of high-grade lead ore was found in Missouri, much of it within the Mark Twain National Forest. The discovery now produces more than half of the nation's lead. Thus, the eastern national forests are endowed not only with known deposits of minerals, but with the potential for discoveries as yet unimagined.

A quick survey of the eastern national forests reveals the following potential mineral "hot spots" in addition to those mentioned:[74]

Superior National Forest (Minnesota)
A proposed land exchange with the Inland Steel Corporation would allow 2,920 acres of forest land to be used as a dumping ground for taconite (iron ore) waste. Elsewhere in the forest, the International Nickel Company seeks to develop a 4,500-acre leased site into an open pit copper-nickel mine. The pit could eventually be over one mile long, three-quarters of a mile wide, and 1,000 feet deep. A nearby river, which could be poisoned with leachate from the tailings, flows into the Boundary Waters Canoe Area.

There are many mining claims within the Canoe Area itself, and copper and nickel prospecting has been proposed there.

George Washington National Forest (Virginia)
Gas drilling is being undertaken by the Washington Gas Light Company, a utility company seeking to expand its "captive" supplies.

Homochitto National Forest (Mississippi)
Production of oil and gas is underway on both private and federal lands. Equipment failures have caused stream pollution and loss of timber, in some cases making the land sterile for several years.

Daniel Boone National Forest (Kentucky)
Some 400,000 acres of coal-bearing lands are subject to privately owned mineral rights. Long considered only marginal to mine, they now are much in demand.

Wayne National Forest (Ohio)
With only 20 percent of the land in federal ownership and only about a third of that with federal ownership of the subsurface, there is considerable oil, gas, and strip-mining activity. It is estimated that mining could disturb 70 percent of the area within the forest.

Kisatchie National Forest (Louisiana)
There are known deposits of oil and gas, salt, low-grade iron ore, and sand and gravel on the forest. Reports a district ranger: "Presently, there is an active search for additional oil and gas; and from preliminary indications, we expect oil and gas operations to be intensive in the near future."

Jefferson National Forest (Virginia)
Of 85,000 acres on the Clinch Ranger District, 55,000 have privately held mineral rights. Approximately 20 coal-prospecting permits for core drilling have been issued on national forest land with reserved mineral rights.

Ouachita National Forest (Arkansas)
Considerable potential for coal leasing exists.

The likelihood that any or all of these known minerals will ever be developed is dependent on future prices, which are themselves to some extent the result of government policy. Future limits on oil imports, national policy on food and fertilizer export, and the possibility of federal strip-mining legislation are all relevant in determining what the demand for these resources might be. Clearly, though, even the possibility that privately held mineral rights will have some future use endows them with considerable value and makes it expensive for the Forest Service to purchase them from their present owners.

PRESERVATION AS A USE

The eastern national forests contain a number of natural features, including various plants and animals, that deserve to be preserved[75] not just for our current use, but for their own sake and for the sake of coming generations. The Presidential Range and Chattooga River; the timber wolf and Kirtland's Warbler; the remaining virgin stands of white pine and eastern hemlock—all have a value that is beyond their "usefulness" even to those who merely wish to view or study them.

Perhaps the greatest benefit of public ownership of the eastern national forests has not been their cumulative production of goods and services (although these have been great), but the fact that their productivity and quality have been preserved and enhanced. On one hand, public ownership has prevented private development of these forest lands for the kinds of shoddy recreational and commercial facilities that have been so destructive to other amenity-rich mountain and forest locales. On the other hand, public investment and resource protection have significantly improved the condition of the timber now standing on the land and have restored wildlife and scenic quality to land that had previously been abused. In this sense, preservation of the stock of forest resources demands equal billing with the useful products and services that flow from that stock yearly.

OTHER USES

Over the years, eastern national forests have been used as convenient, yet out-of-the-way, locations for various land-intensive federal projects. One of these is military use. Louisiana's Kisatchie National Forest contains an Army base, an Air Force base, and a Louisiana National Guard encampment. Past military use has affected the management of each of the six ranger districts and present use affects multiple-use management of 20 percent of the land.[76] Both the Kisatchie and Ocala have active military bombing ranges. An entire 350,000-acre forest in Florida, the Choctawatchee, was taken over by the military in World War II and is now Eglin Air Force Base. (Much of its land is managed for forest use by Defense Department foresters).

Large public reservoirs have been constructed on several national forests. In Texas, about 60,000 acres of forest lie under the waters of the Sam Rayburn, Toledo Bend, and Lake Conroe reservoirs. A score of small flood-control dams have been built on Virginia's George Washington National Forest, as has a small pumped storage reservoir built by a private utility.

National forest land is also used for pipeline rights-of-way, television towers, electric transmission and telephone lines, highways, and municipal landfills. In a few cases, towns within the forest boundaries have expanded up to the edges of the federal lands, and the Forest Service has been approached with proposals to trade land to them for future expansion.

The eastern national forests, whose boundaries encompass some 47,000,000 acres of land, are likely to face an incredible diversity of pressures for new uses in the decades ahead. Some future decision makers may regard these forests as potential locations for facilities best kept out of built-up areas. National forest lands may be sought after as sites for transportation and communication facilities—high-voltage transmission lines, airports, microwave relay towers, television antennas, highways, or perhaps for structures or rights-of-way for technologies as yet uninvented. Other lands may be sought for waste-disposal sites, perhaps as places to bury common garbage or the sludge from sewage-treatment plants, or to accommodate the land spraying of nutrient-rich liquid effluent.

Because of their size and relative isolation, these lands also may be desired as sites for so-called "energy parks," combining large numbers of electric-generating units, either fossil or nuclear fueled. The concentrated air pollution produced or possible radiation danger would clearly require that such facilities be located far from population centers.

If more speculative energy systems are developed, the eastern national forests could be called upon to provide sites for huge banks of solar cells as an element in a crash national energy effort. Alternatively, the forests might provide great quantities of wood for methanol production, which would involve an entirely different kind of timber management. One experimental methanol system calls for the use of species such as poplar, alder, and willow, planted at densities somewhere between pulpwood and a cornfield, and cut when two inches in diameter.

In the future, some communities may rely on the eastern national forests to provide a wooded background to a pattern of dispersed living.[77] As privately held inholdings and adjacent lands are built upon, large sections of the forests may take on functions little different than those of large urban parks. Laced with access roads and utility lines, they would lose both their "wild" character and much of their usefulness for timber supply. In such cases, the justification for continued federal ownership would have to be questioned; the parts of the forest most affected might logically be transferred to state or local government ownership.

The Users of the Forest

The statistics by which forest uses are measured provide only a shadowy picture of the people who enjoy those "visitor days" of forest recreation, or harvest and use those "cubic feet" of timber products. Perhaps the most useful way to describe them is according to whether they use the national forest to produce income ("forest producers") or for their own consumption ("forest consumers"). One person, of course, can be both producer and consumer, relating to one forest policy issue as a producer and to another as a consumer. Other characteristics of forest users also should be identified: Are they located near the forest or far away? Do they pay market prices for the forest resources or enjoy a public subsidy?

FOREST PRODUCERS

Most obvious among those who use the eastern national forests to produce income are those who are involved with timber. These include the logging company which purchases stumpage; the sawmill that cuts the logs into lumber or peels them for veneer; and the building contractor or furniture plant that uses the wood in a final product. Some of these producers are located near the forest, perhaps even in a town within forest boundaries. Others are far away, often in metropolitan areas.

There is no easy way to determine how dependent these users are on wood supplies that originate in the eastern national forests. The Forest Service estimates roughly that each dollar of stumpage value eventually results in wood products worth $25.[78] On this basis, the $33 million in annual eastern national forest timber sales could generate $825 million in products.

But this calculation would considerably overestimate the economic contribution of the forests, for it assigns to the forest itself the value created by all the labor and capital expended in manufacturing wood products throughout the stages of production. It also assumes there are no substitutes for this particular timber in the production process. In fact, there are many substitutes for east-

The Allen-Rogers Company has three sawmills, one in New York and two in New Hampshire. The New Hampshire mills get about 25 percent of their logs from the national forest; it accounts for about 5 to 10 percent of our total log supply.

The future supply of timber will be determined by land-use laws and the availability of timber from the national forest. If cutting were stopped on the national forest, competition would be keener. We would still be here and operating, but many smaller companies would be hurt. As more and more forest land is locked up, the price of timber goes up because of the decreased supply. Ultimately, the consumer would pay the bill, because the price of the products we sell the consumer would be higher. I can't state it any simpler than that.

RICHARD H. BURT, Vice President/Works Manager, Allen-Rogers Corporation
Laconia, New Hampshire

---

ern national forest timber, with the possibilities for substitution increasing the farther one gets from the individual forest. Rough logs typically are shipped no more than 100 miles from forest to sawmill. Thus local mills, if deprived of eastern national forest timber, would have to find private supplies of logs somewhere within this radius. Finished and semifinished products, on the other hand, often are shipped great distances. It is not uncommon, for example, for lumber used in buildings in the northeastern United States to have originated in Oregon or Washington.

In trying to estimate how much economic activity actually depends on eastern national forests, it is necessary to look, then, not at all users of timber, but at those who use rough logs.[79] Basically, these are sawmills, pallet and container producers, and pulp and paper mills. In 1967, these log-using industries employed 500,000 people in the eastern and southern states. Since the eastern national forests contributed 2 to 3 percent of wood production in these regions, it can be assumed that they supported 10,000 to 15,000 of these jobs.[80] Without continued log sales from the eastern national forests, a certain proportion of these jobs would be eliminated, the exact number depending on each mill's ability to acquire an alternate supply of timber. Producers further down the line, such as homebuilders, simply would obtain their lumber from other sources and—except for a possible small cost difference—would not be affected by the substitution.

The White Mountain National Forest accounts for only about five or six percent of our timber requirements for the Brown Company pulp mill here at Berlin, New Hampshire. It seems like a small percentage, but it is significant. We have no large single source of timber, even though we own a lot of land. Rather, we have to rely on many small sources. We're concerned about losing any source.

While we have no real worry now, there has been environmentalist pressure on all national forests over the harvesting of timber. We can't help but conjecture that there might be a movement in the future to reduce the harvest of timber on the national forests. The White Mountain National Forest is a recreational forest; the emphasis on recreation could reach the point where harvesting of timber could be appreciably reduced. Fortunately, most of the people who are interested in the White Mountain National Forest seem to want to keep it a forest and not turn it into a national park, which was proposed at one time. We're pleased, of course, in the support for retaining the national forest.

People who use the national forest should be exposed to the harvesting of timber—and realize that the forest products they use day in and day out have to come from somewhere. The national forest should serve as an educational tool. People shouldn't feel badly when they see a stump. Instead they should ask, "What was that tree used for?"

I realize there are 30 million people in megalopolis who need places to recreate, but if you ignore industry—and in this area I mean the timber industry—I shudder to think of the type of local economy you would create. We'd have nothing but chambermaids and handymen. I think the people who live here are entitled to more than that. The forest should not

be just a playground for people who have elected to live in megalopolis. People who earn enough money to go out and recreate usually earn it through industry—some raw material somewhere being turned into wealth, being distributed and redistributed and maintained—and there's a few bucks left over for recreation. Recreation doesn't create wealth; basic industries, like timber, have to be given some consideration.
JOHN H. BORK, General Manager-Woodlands, Brown Company
Berlin, New Hampshire

Also dependent on the forests for income are those who sell goods and services to forest recreationists. These include businesses near forests, such as gas stations, restaurants, hotels, and campgrounds, as well as those quite removed, such as manufacturers of camping equipment, fishing tackle, and boats. Again, it is difficult to say what proportion of their sales depends on the existence of the eastern national forests as a source of outdoor recreation. It is quite possible, though, that the economic impact of forest recreation is comparable to that of the forests' timber production. For example, in 1974, hunting and fishing accounted for 11,000,000 visitor days of eastern national forest use. Nationally the average hunter is estimated to spend $10.61 for each day of hunting activity; the average freshwater fisherman, $6.30.[81] Using these figures, direct expenditures on eastern national forest hunting and fishing approximate more than $90 million. Perhaps $20 million of this is spent in communities near the forests.[82] This money, circulating in the community, helps support a whole group of producers, dependent on forest recreation for their livelihood.

The governments of the counties that contain eastern national forests also depend on them for income. Some 371 counties include eastern national forest land within their borders; 338 counties contain more than 1,000 acres of it.[83] These counties levy no property taxes on national forest land, but have received 25 percent of the annual forest receipts. Under legislation enacted by Congress in 1976, counties now have the option of receiving a flat 75 cents per acre for land in national forests. Many poor eastern counties will realize significant payment increases as a result of this legislation. (See pages 225-231.) In 1974, payments to counties in the East amounted to $11,200,000.[84] For most counties, this revenue represents a small proportion of total receipts, although one South Carolina county received over $500,000 ($83 per capita). Localities also receive revenues raised by taxes on purchases in nearby communities by forest recreationists. On the expenditure side, local governments provide few or no public services to the national forest lands.[85]

Local businesses and governments also benefit financially from Forest Service operations, which generate hefty payrolls. In 1975, Forest Service management expenditures in the eastern states containing national forests totaled $62 million, much of which was spent locally.[86]

Also reaping economic benefit from national forests are some owners of inholdings and adjoining lands. If they use their land for timber, these neighbors of the national forest receive little or no special benefit from their location. If, however, they use their land for recreation or subdivide it into building lots, they enjoy, in effect, a federally provided greenbelt adjoining their property.

There is a growing realization that growth is not always wonderful. If a 100-acre parcel is subdivided and built full of A-frame houses and septic tanks, then it could actually be a net tax loss to the town. On the other hand, the land might be bought by the Forest Service and taken off the tax rolls. Then the town will get 25 percent of gross Vermont National Forest revenue, based on that town's percentage of total national forest acreage.

BRENDAN WHITTAKER, Chief, Information and Education, Agency of Environmental Conservation
Montpelier, Vermont

Hunters, for example, have easy access to game on federal lands. Owners of second homes have neighboring land that, in all probability, will never be built upon. It is not uncommon to find newspaper advertisements for private lands with such phrases as "adjoining a national forest" or "surrounded by national forest."

One Forest Service official draws a map of an inholding and describes how the government comes under pressure to preserve a buffer strip of woodland around it to protect the aesthetic value of the inholding. "The public," he says, "is providing a buffer strip for private land and by preserving the buffer, the public is increasing the value of the inholding." [87]

FOREST CONSUMERS

The most visible consumers of the services of the eastern national forests are the forest recreationists, whose use may be as extended and premeditated as a week-long camping vacation or as casual as a drive on a forest highway on the way to another place. According to a national survey, people drive considerable distances to participate in outdoor recreational activities.[88] About 95 percent of vacation and overnight trips, 90 percent of day outings, and more than 50 percent of short (up to four hours) trips taken for outdoor recreation involve distances that would typically take the participant into a county other than his own.[89] A great deal of the recreational demand for national forests thus comes from persons who live outside the immediate forest environs, including many who live beyond the zone in which most "forest producers" are located. About three-quarters of national forest users, in fact, live in metropolitan areas.[90]

The forest recreationists pay no fees—or only minimal ones—for use of the forest resource, although their indirect outlay may be substantial. Under a 1974 law, the Forest Service charges fees only for the most highly developed of its facilities, which in the East means that charges are made only on about half of the developed campgrounds and one-quarter of the swimming beaches.[91] Other forest uses, including driving scenic highways, hunting and fishing (except for state licenses, where applicable), camping on primitive sites, and use of interpretive centers are free to the consumer. User charges play such a small role in the eastern national forests that in 1974 the government collected only slightly more than $1,000,000 in user fees, although these lands sustained 44,000,000 visitor days of use.[92]

TABLE 5
RELATIVE PREFERENCE OF VARIOUS SOCIOECONOMIC GROUPS FOR OUTDOOR RECREATIONAL ACTIVITIES

M = males   U = under 25   U = urban   L = less than $8,000   H = high school   W = white
F = females O = 25 and over   R = rural   M = $8,000- $14,999   or less   N = nonwhite
                                              H = $15,000 and over   C = college

Activity	Sex	Age	Residence	Income	Education	Race
Camping in remote or wilderness areas	M	U	R	M&H	*	W
Camping in developed camp grounds	M	*	R	M&H	C	W
Hunting	M	U	R	*	H	W
Fishing	M	*	R	*	H	W
Riding motorcycles off the road	M	U	R	*	*	W
Wildlife and bird photography	M	U	U	H	C	W
Bird watching	*	O	U	*	C	W
Hiking with pack; mountain/rock climbing	M	U	U	M&H	C	W
Nature walks	F	U	U	M&H	C	W
Swimming outdoors	F	U	U	H	C	W
Sightseeing	F	*	U	M&H	*	W
Picnicking	F	*	R	*	*	W
Driving for pleasure	F	*	R	M&H	*	W

* Data do not reveal a definitive preference.

NOTE: The symbols in the table should be interpreted for all activities as in the following example of hikers. The percentage of males who hike is higher than the percentage of females who hike, the percentage of those under 25 who hike is higher than the percentage of those over 25 who hike, the percentage of urban residents who hike is higher than the percentage of rural residents who hike, etc.

SOURCE: U.S. Dept. of the Interior, Bureau of Outdoor Recreation, *Outdoor Recreation: A Legacy for America* (Washington: GPO, 1973), Appendix A, p. 7.

A typical national forest user, according to profile data, is a white male, less than 44 years old, residing in a metropolitan area, enjoying an average or above-average income.[93] Obviously, this is only a statistical construct; some forest recreationists are black, female, or old. There are also some differences that become apparent only when the type of use is considered (Table 5). Wilderness camping, for example, is more likely to appeal to younger males, while driving and picnicking gain relatively more participation from females and show no variation with age.

Consumers of timber products are somewhat more difficult to characterize. Almost everyone, regardless of location, age, or income, uses wood products and paper. Consumption of these articles seems to be proportional to income, with higher income people devoting a slightly smaller proportion of their total expenditures to housing than do the less affluent, but spending a somewhat higher proportion on home furnishings and reading material.[94] Unlike forest recreationists, users of national forest timber pay approximately market prices for what they consume.[95]

The future users of the eastern national forests are, broadly speaking, likely to fall into the same groups as they do now. Such factors as increased population in forest-area communities, technological change in the timber industry, and changes in public tastes in recreation could alter their relative numbers somewhat, but it is most unlikely that any of the groups now represented will not be important a generation from now.

MANY DEMANDS, MANY DEMANDERS

Economists say that human wants, unless restrained by high prices or other limitations, are nearly infinite. Forests and their products are no exception to this rule. Almost everyone would like more from the eastern national forests—and at the same price (in money or in convenience) as the last supply. The timber company and the homebuilder would like more timber; the camper, more campsites; the hunter, more deer and turkey; the wilderness hiker, more solitude.

As the demands of these users increase, the resource cannot continue to meet them all. Development of long-range objectives that both realize the special qualities of the eastern national forests and permanently protect the forest resource provides a framework for establishing priorities among the various demands and uses.

REFERENCES

Chapter II

1. Gifford Pinchot, *Breaking New Ground* (New York: Harcourt, Brace and Company, 1946), p. 262.
2. U.S. Forest Service, *The Outlook for Timber in the United States* (Washington: GPO, 1974), Appendix tables. (Hereafter cited as *Outlook for Timber.*)
3. Ibid.
4. A pallet is a portable platform used with a forklift to move boxes or stacks of materials.
5. U.S. Forest Service, *Outlook for Timber*, Appendix tables.
6. Ibid.
7. Ibid.
8. Ibid.
9. Data are from U.S. Forest Service, program development and budget staff. Receipts are net of K-V funds, but not of payments to local governments.
10. A notable exception is revenue from lead mines on the Mark Twain National Forest in Missouri, which bring in about $6,000,000 yearly.
11. U.S. Forest Service, program development and budget staff.
12. For example, the funds "invested" in trees long ago could have been used to pay off a portion of the national debt, with resulting savings in interest costs.
13. Data are from U.S. Forest Service, timber management staff.
14. National Forest Management Act, P.L. 94-588.
15. Proportionately more land is disturbed in southern forests than in the North, but growth of new trees, particularly in pine stands, is also faster.
16. "It is the policy to manage each working circle for the production of crops of sawtimber size and quality from all suitable forest types and sites, unless exceptions are approved for a particular working circle." *Forest Service Manual*, title 2400. Some areas of aspen, jack pine, and sand pine are managed for pulpwood.
17. Ironically, effective fire control may have played a large part in this, for pine forests seem to thrive under conditions of frequent, low-intensity fires.
18. "Are We Managing Our Forests to Death?" *ENFO*, newsletter of the Florida Conservation Foundation, February 1975.
19. Interview, September 1976.
20. "Herbicide Use in Ozark Forests Challenged," *New York Times*, July 14, 1975.
21. *Forest Service Manual*, Title 2410.3 (May 1972).
22. Projections of Leonard L. Fischman, "Future Demand for U.S. Forest Resources," in Marion Clawson, ed., *Forest Policy for the Future* (Washington: Resources for the Future, 1974).
23. A. B. Makhijani and A. J. Lichtenberg, "Energy and Wellbeing," *Environment*, Vol. 14, no. 5 (June 1972), pp. 10-18, cited in Jerome Saeman, "Solving Resource and Environment Problems by the More Efficient Utilization of Timber," in *Report of the President's Panel on Timber and the Environment*, pp. 354-55.
24. A comparison of direct and indirect purchases from the coal industry and the petroleum and natural gas industry reveals the following relationship (per

dollar of sales):

Lumber and wood products	$2.38	Heating, plumbing, and	
Paper and allied products	2.24	structural steel	$1.28
Plastics and synthetics	6.19	Primary aluminum	1.28

SOURCE: "Input-Output Structure of the U.S. Economy, 1967," *Survey of Current Business*, February 1974, pp. 24-56. Note that energy requirements of the aluminum industry are understated to the extent that the electricity used is generated by water power.

25. William A. Duerr, *TIMBER!: Problem, Prospect, Policy* (Ames, Iowa: Iowa State University Press, 1973), p. xiv.
26. The 1968-70 run-up led to the proposed Timber Supply Act, while the 1972-73 boom resulted in the President's Advisory Panel on Timber and the Environment.
27. Developments in the technology of wood use are among the imponderables in estimating future demand for particular types of wood. For example, the recent approval in building codes of the use of aspen for studs has substantially increased demand for this product of the Lake States forests. Previously, the species had been considered of little value.
28. Southern Forest Resource Council, *The South's Third Forest* (Atlanta: S.F.R.C., 1969).
29. This will happen regardless of changes in national forest timber policy.
30. Southern Forest Resource Council, *The South's Third Forest*, p. 40.
31. U.S. Forest Service, *Guide for Managing the National Forests in the Ozark Highlands* (Atlanta and Milwaukee: U.S.F.S., 1974), p. 22.
32. Interview with Jack Muench, National Forest Products Association, November 1975.
33. *Southern Lumberman*, June 15, 1975, p. 4.
34. U.S. Forest Service, *A Recommended Renewable Resource Program* (Washington: U.S.F.S., 1975), p. 282.
35. U.S. President's Panel on Timber and the Environment, *Report*, p. 84. See also, Marion Clawson, "The National Forests—A Great National Asset is Poorly Managed and Unproductive," *Science*, Vol. 191 (February 1976), pp. 762-67.
36. U.S. Forest Service, *Outlook for Timber*, Appendix Table 5.
37. Obviously, investment potential depends on the kind of species a site will support as well as the rate at which it will grow trees. For example, Vaux calculates that it is unprofitable to grow true firs even on some very good California sites. See Henry J. Vaux, "How Much Land Do We Need for Timber Growing?" *Journal of Forestry*, Vol. 71, No. 7 (July 1973). Many eastern national forest sites, especially in the South and Lake States, are capable of growing desirable species and, because of their lack of slopes, are easy to cut.
38. U.S. Forest Service, *Report of the Chief, 1974* (Washington: U.S.F.S., 1974), pp. 18-19. A "visitor day" is defined as 12 hours of use.
39. The national park figure excludes visits to the large number of new parks created during this period.
40. A congressionally mandated experiment with admission charges to national forests and other public lands began in 1965. Its applicability to the forests was repealed in 1972. Little revenue was raised while it was in effect.
41. Data provided by U.S. Forest Service, recreation staff.
42. U.S. Forest Service, *Guide for Managing the National Forests in the Appa-*

lachians, 2nd Edition (Atlanta: U.S.F.S., 1973), p. 23.

43. U.S. Forest Service, *Guide for Managing the National Forests in the Coastal Plains* (Atlanta: U.S.F.S., 1974), p. 13. See also *Guide for Managing the National Forests in the Ozark Highlands*, p. 19.
44. Testimony of Associate Chief Rexford Resler to U.S. Senate Interior Committee, *Outdoor Recreation*, 94th Congress, 1st Session, Feb. 5, 1975 (Washington: GPO, 1975), p. 94.
45. Revenue to the government from the 10 ski permits outstanding in New England in 1971 was less than $8,000. See U.S. Forest Service, *New England Area Guide* (Milwaukee: U.S.F.S., 1972), pp. 2-14.
46. Glen Robinson, *The Forest Service: A Study in Public Land Management* (Baltimore: Johns Hopkins Press for Resources for the Future, 1974), p. 147, n. 6.
47. Testimony before U.S. Senate Interior Committee, *Eastern Wilderness Areas*, February 21, 1973 (Washington: GPO, 1973), p. 23.
48. Testimony of Ernest Dickerman, director of field services, Eastern Region, Wilderness Society, in Ibid., p. 47.
49. There are other wilderness areas in the eastern states on lands managed by public agencies other than the Forest Service.
50. For a list of areas, see "The Wilderness System," *The Living Wilderness* (Winter 1974-75), pp. 38-47.
51. See H. E. Wright, Jr. and Jonathan Ela, "Cutting Up the Boundary Waters," *Sierra Club Bulletin* (May 1974), pp. 24-28.
52. Interview with Tom Roederer, U.S.F.S. recreation staff, August 14, 1975.
53. Malcolm Baldwin and Dan Stoddard, Jr., *The Off-Road Vehicle and Environmental Quality* (Washington: The Conservation Foundation, 1973).
54. U.S. Forest Service, *Nation's Renewable Resources: An Assessment* (Washington: GPO, 1976), p. 102.
55. U.S. Congress, House Committee on Interior and Insular Affairs, *Eastern Wilderness Areas*, 93rd Congress, 1st Session, 1973, part 3, pp. 132-34.
56. Personal communication, February 1976.
57. American Society of Planning Officials, et al., *Subdividing Rural America* (Washington: GPO, 1976), Executive Summary, Appendix C.
58. U.S. Forest Service, Eastern Region, *The Fairest One of All* (Milwaukee: U.S.F.S., 1973).
59. See, for example, U.S. Forest Service, *Report of the Chief*, 1974, p. 28.
60. Interview with Donald Strode, U.S. Forest Service wildlife staff, August 15, 1975.
61. For example, says an aide to West Virginia Senator Jennings Randolph, "Cranberry River in Nicholas County, West Virginia, ran muddy for seven weeks following a massive clearcut on the Gauley ranger district [in the Monongahela National Forest]. Roads there have not been built to Forest Service specifications." Interview with Bill Davis, August 1976.
62. Interview with Ed Johnson, U.S. Forest Service watershed management staff, June 26, 1975.
63. According to the U.S. Water Resources Council, year 2000 water demands for most basins in the East and Southeast are projected to be well below the usual yearly runoff. The major exceptions are South Florida and southern Lake Michigan, neither of which is much affected by the national forests. Data

cited in U.S. Forest Service, *Nation's Renewable Resources: An Assessment*, pp. 314-18.
64. U.S. Forest Service, *The Nation's Range Resources: A Forest Range Environmental Study* (Washington: GPO, 1972), Table 57, Alternative 19.
65. In past decades, the continuing increase in beef cattle has been offset by a decline in the number (although not in the productivity) of dairy cows.
66. Interview with Melvin Bellinger, U.S. Forest Service range staff, August 1975.
67. U.S. Forest Service, *Nation's Renewable Resources: An Assessment*, p. 187.
68. U.S. Forest Service, *Recommended Renewable Resource Program*, pp. 227-67.
69. Letter of John R. Castles (former director of timber management, Eastern Region), June 22, 1976; interview with Edward Cliff (former chief of the Forest Service), July 1976; interview with Dr. David Smith, Yale University School of Forestry and Environmental Studies, October 1976.
70. Much of this discussion draws on Margaret B. Coon, *Mining in the Eastern National Forests* (Washington: The Conservation Foundation, 1975). It is also based on the various forest plans for the specific areas cited and inquiries to forest supervisors.
71. *Mineral Considerations in Weeks Law Purchases and Exchanges*, report to the National Forest Reservation Commission, January 1972, p. 3.
72. Cited in Monongahela National Forest, *Coal Mining: Its Situation and Its Management* (1970).
73. *Mineral Considerations in Weeks Law Purchases*, p. 6.
74. See note 70.
75. In many cases, particularly for wildlife, preservation means not just protection but active management as well.
76. U.S. Forest Service, *Guide for Managing the National Forests in the Coastal Plains*, p. 40.
77. Discussions of how improved communication technologies might affect spatial patterns may be found in Brian J. L. Berry, "The Geography of the U.S. in the Year 2000," *Ekistics*, Vol. XXXII, No. 174 (1970), pp. 339-51; and in Peter C. Goldmark, "Communication and the Community," *Scientific American*, September 1972.
78. This is in accord with the independent estimate that southern forest stumpage valued at $500 million added $14 billion to the national economy. See Southern Forest Resource Council, *The South's Third Forest*, p. 34.
79. Industries considered were those making significant direct purchases from the "logging camp and lumber contractor" sector in the 1963 input-output table.
80. Wages in these industries are close to the national average for all sectors, with paper being somewhat higher paying, logging and lumber somewhat lower.
81. U.S. Fish and Wildlife Service, *National Survey of Fishing and Hunting, 1970* (Washington: GPO, 1972), pp. 5 and 9.
82. Based on percentages of total expenditures going to "food, lodging and transportation" and "bait, guides and other." Assumes that one-half of these expenditures are made in communities near the destination forest.
83. There are 13 additional counties that are within an eastern national forest boundary but in which no land has yet been purchased by the Forest Service.
84. U.S. Forest Service data.
85. A 1971 law provides for cooperative agreements between the Forest Service

and local governments for law enforcement, but the costs of this service are reimbursed by the Federal Government.
86. Includes salaries, building rent, purchases of supplies, etc. See *Department of the Interior and Related Agencies Appropriations for 1976,* U.S. House of Representatives, Committee on Appropriations, 95th Congress, 1st Session (Washington: GPO, 1975), p. 448.
87. Interview with Kenneth Scholz, U.S. Forest Service lands staff, April 25, 1975.
88. U.S. Department of the Interior, Bureau of Outdoor Recreation, *Outdoor Recreation: A Legacy for America* (Washington: GPO, 1973), Appendix A, p. 26. Because this survey was based on self-reporting rather than actual count, it may somewhat understate the proportion of short trips, which would be less likely to be recalled by the participant.
89. This rough calculation assumes the participant lives at the center of a county of average size.
90. U.S. Forest Service, *Nation's Renewable Resources: An Assessment,* p. 58.
91. Data supplied by Gordon Sanford and John Tucker, U.S. Forest Service recreation staff.
92. See notes 38 and 41.
93. U.S. Forest Service, *Nation's Renewable Resources: An Assessment,* p. 58.
94. U.S. Bureau of Labor Statistics, *Survey of Consumer Expenditures, 1960-61* (Washington: GPO, 1964).
95. This is only approximate because (1) national forest timber investment and sales decisions are not made on strictly economic grounds; and (2) purchasers of national forest timber must meet environmental standards that are higher than those applying to cuts on nonfederal land.

CHAPTER III

objectives, opportunities, and incentives

In the six decades since they were established, the eastern national forests have been able to meet the demands placed on them with surprisingly few conflicts. To be sure, there has been public controversy over such issues as clearcutting on the Monongahela National Forest, phosphate mining on the Osceola, regulation of snowmobile use, and the size and location of wilderness areas. But to date, levels of demand for forest uses have been such that multiple-use management has meant satisfying most of the users most of the time. However, the strong and conflicting demands increasingly being made upon this limited resource base, both in traditional use areas and in new ones, suggest that this relatively happy situation cannot long continue.

In the future, forest-management policy is likely to involve many more painful choices and decisions about which of the various current and potential demands on these lands should be accommodated. Any realistic and meaningful long-term goals for this land system should emphasize those special — and sometimes unique — opportunities for public benefit that these lands offer. To determine the special potentials of the eastern national forests, it is useful to look at the role of all eastern forest land — public and private — in the national context, and then to compare the characteristics of eastern national forests with those of eastern private lands.

Additionally, any decision about future uses of the eastern national forests is necessarily constrained by what these lands are capable of producing. These 24,000,000 acres of public land contain significant amounts of timber and considerable potential for growing more — yet their timber stocks are small compared with that on federal land in the Pacific Northwest or growing on private land in the South. The eastern national forests are a great recreational resource, accessible to the country's most densely populated urban regions — yet they amount to only 6 percent of the forested land in the eastern states. They include a number of tracts of land of wilderness quality — but these are minor indeed when compared with the wilderness that exists on the expanses of public land in the West. The eastern national forests can meet many of society's demands, but their ability to do so is rather strictly constrained by their limited resource base.

The development of goals for these forests should consider needs and demands far beyond those of the current generation. This means that where the resources of the forests are renewable — such as timber, wildlife, or water — they must be used in ways that ensure their equal availability in the future. The Congress has recognized this responsibility by requiring that national forests be managed according to the principle of sustained yield. Where resources — such as minerals — are not renewable, they should be used conservatively, recognizing that new technologies may not unendingly appear to furnish new low-cost supplies. Finally, exceptional caution should be exercised in making irreversible commitments of resources that cannot be replaced by human action. This applies particularly to opening natural areas to their first development, to elimination of species variety, and to changes in landscape features.[1]

The responsibility to conserve resources for future generations applies, of course, to private forests as well as public. Realistically, however, public land managers should strive to achieve particularly high standards, in part to provide an example to the private sector and also to provide a cushion against society's mistakes.

Within these limits, many of the demands that people wish to make of forests are quite legitimate. Some of the alternatives — for instance, using more steel and less lumber in construction or emphasizing shoreline rather than forest recreation — involve far greater potential for environmental damage.

What, then, are reasonable objectives for these lands in both national and regional contexts?

FORESTS—EAST AND WEST

Demands for most forest products, whether timber, recreation, or open space, probably will be higher in the future than they are today (see Chapter II). Some of the heaviest demands, particularly for near-term timber production and for the preservation of very large tracts of wilderness land, undoubtedly will be made on the larger forests of the western states and Alaska. But many will be made in the East, both on public and private forests. Those enumerated in Chapter II include demands for more southern softwood lumber and pulp, for hardwood fiber from heretofore worthless sizes and species, for habitat for forest wildlife, and for a variety of highly specialized forms of outdoor recreation.

Since forests east of the Rockies contain nearly 90 percent of the nation's hardwood growing stock, there clearly is not much opportunity for supplying hardwood lumber without logging somewhere in the East. For softwoods, the warm climate and abundant rainfall of the southern states provide considerable opportunity for intensive forestry. Indeed, in recent years, private timber companies have purchased millions of acres in the South for softwood plantations, many of them planted with genetically superior trees.

For two centuries, the lumber industry worked its way across the country, cutting the primeval stands of New England and New York, then the Lake States, the South, and currently, the Pacific Northwest. Now for the first time, really significant portions of the nation's timber are starting to come from second- and third-growth stands, some of them artificially planted. In this situation, attention is increasingly turning to the potential productivity of forest land, rather than simply to the current value of its trees.

From this perspective, the eastern states look extremely significant as potential timber producers. Of all of the nation's best timberland (land capable of producing more than 85 cubic feet per acre per year), more than two-thirds lies in the eastern states. About 18 percent of this is forest industry land; another 3 percent is on the eastern national forests. Over 70 percent is private, nonindustrial land, much of it currently given little or no attention for timber production by its owners.

Perhaps the major determinant of the rate at which timber cutting shifts from the West to the East and South is what the Forest Service decides to do with the tremendous stock of mature trees now standing on national forests in the western states. These trees account for nearly half of the nation's entire standing stock of softwood sawtimber. Most are in virgin or "old growth" stands that have never been cut and are now being liquidated very slowly.

Each time in recent years that lumber prices have risen sharply, there have been calls for scheduling faster rates of cut of these old trees. Some argue that maintaining so large an inventory of mature trees is wasteful and that they should be replaced with a younger, faster-growing forest. Opponents point out that to cut these trees, roads would have to be pushed into remote areas, sometimes over steep slopes susceptible to erosion or into potential wilderness tracts. It would take generations to reproduce the present mature forest.

Thus future policies for eastern and western timberlands are interrelated. Greater immediate production from public lands in the West can be emphasized, increasing the rate of cut of the old growth stands. Or the eastern and southern states, with their high fertility and existing road system, can assume a greater role as sources of future softwood supply. In view of such a choice, it seems preferable that intensive forestry increasingly be practiced on the naturally fertile eastern and southern lands, and that liquidation of old growth forests in the West proceed very gradually.

As recreational lands, eastern forests generally are higher in accessibility and lower in scenic quality than those in the West. Although there certainly are exceptions, the well-watered forests of the East seem able to tolerate intensive use better than many western forests. Says one eastern forest supervisor, who has also worked in the West:

> I think of the Maroon Bells-Snowmass Wilderness Area in the Rockies, where standing on a ridge one day I could look down one side of the ridge and see people miles away coming toward me, and look down the other side of the ridge for miles and see other people. The area actually seemed crowded, even though there was only a handful of us in a huge area of land.
>
> Here in the East I sat down to have lunch one day on the Appalachian Trail, and just around the bend a few hundred yards was another party also having lunch—I didn't know they were there. The nature of the terrain and soil, the density of the vegetation, the productive potential of the areas give the eastern national forests a great capacity.

Camping, fishing, hunting, hiking, and other recreational uses of eastern forests can be enjoyed by those who live a day's drive or less from the forest. Therefore, in only a small percentage of cases (such as recreation undertaken during lengthy vacations) is there much opportunity to transfer recreation uses from eastern to western forests.

Eastern forest ecosystems differ considerably from those elsewhere in the country. No western substitutes exist either for the common forms of eastern forest life (the oak-hickory forest and its associated animals and birds) or for the rare or even unique ones (pockets of virgin white pines skipped over by colonial lumbermen). It is difficult to say how much of each plant or animal community or type of scenery is "enough" for society, but many can be provided only on eastern forest land.

It would be neither easy nor desirable to try to shift from eastern to western lands future timber or recreation demands or demands to preserve regional open space and unique ecosystems. There are indications, in fact, that the national interest—both economic and environmental—would be better served if eastern forests, both public and private, supplied a somewhat greater share of each.

SPECIAL QUALITIES OF THE EASTERN NATIONAL FORESTS

Clearly, the great bulk of demands on eastern forested lands must be met by lands other than the 6 percent lying within eastern national forests. Their small area, however, obscures the fact that the eastern national forests have some very special qualities—features that distinguish them from the rest of the land in the eastern half of the nation and enable them to provide unique values for society.

One of the most striking attributes of the 50 units that comprise the eastern national forests is their size, in comparison with that of other eastern tracts in single ownership. Thirty-one contain at least 250,000 acres of publicly owned land—making each more than 10 times the size of the District of Columbia.[2] Four forests (Superior, Ouachita, Mark Twain, and Ozark)[3] are larger than a million acres each.

Other single ownerships of this size are quite uncommon in the eastern states. There are somewhat more than 50 forest-industry tracts of more than 250,000 acres, most of them in Maine and the deep South.[4] Fewer than a dozen non-Forest Service public lands in the East contain more than 250,000 acres. These include New York's Adirondack Park (2,300,000 acres of state-owned land), Great Smoky Mountain and Isle Royale National Parks (about 500,000 acres each), Everglades National Park (1,400,000 acres), and state forests in Pennsylvania, Michigan, and Minnesota.[5]

Even with the scattered ownership pattern characteristic of the eastern national forests, possession of large acreages in a well-defined area provides the Forest Service with many opportunities

> Management of Forest Service lands has a different orientation than that of state lands, at least in Massachusetts. For one thing, most national forest lands are in very large tracts— hundreds of thousands of acres— whereas state lands are usually much smaller. In most cases, the federal lands are located in areas remote from the threat of urbanization. They are not managed principally for recreation purposes, but primarily for conservation purposes.
>
> BETTE WOODY, Boston University
> Former Massachusetts Commissioner of Environmental Management
> Boston, Massachusetts

for setting the character of land use over a large area. Large tracts of land in single ownership also increase present management opportunities and preserve future options. Although it is easy to subdivide land, it is difficult indeed to piece a substantial tract together from smaller ones, particularly when some of the small holdings are of building-lot size or have already been built upon. The history of the eastern national forests themselves bears this out. Although they were created out of unwanted lands, in times of low prices and economic hardship, their ownership was never consolidated. But even with their fragmented ownership, the eastern national forests represent a degree of control over very large tracts of land which would be difficult or impossible to create today.

In addition to their size, the eastern national forests are notable for their absence of intensive development. Although relatively few tracts of absolutely pristine wilderness exist in the eastern states, a number of areas remain which the hand of man has touched only lightly. One geographer identified these as "empty areas," eastern land areas of a square mile or more that neither contained occupied structures nor were used for farming or industry.[6] For reasons topographical, climatic, or economic, these places — many of which are forested — were skipped over while the rest of the region was being settled. There are persuasive reasons to retain these lands as open space.

While man may be currently absent from these areas, other life abounds there and they are prime habitats for other species. Though empty of man, they are not empty of value to him.[7]

They are a prime source of watershed, habitat for solitary wildlife, and dispersed recreation. Their existence gives context and form to the developed part of the landscape. While the frantic pace of land development in the last two decades suggests that privately owned "empty" lands are fast disappearing,[8] the national forests, devoted mainly to "natural" uses, remain about as undeveloped as they ever were.

The eastern national forests are also special for their sites of high scenic or natural value. Certainly many areas in eastern national forests are not particularly high in amenity value, just as a great many high-amenity areas exist outside the national forests. But the national forests seem to have more than their share of such places, partly because so many of the high mountain ridges, both in the Southern Appalachians and in New England, lie within forest boundaries. National forests occupy much of the White Mountains in New Hampshire and Maine, the Green Mountains in Vermont, the Allegheny and Cumberland Plateaus, the Blue Ridge Mountains of Virginia and North Carolina, and the Ozark and Ouachita Mountains of Arkansas and Missouri. National forests boast such spectacular areas as Mount Washington, the highest point in the Northeast, and Mount Mitchell, the highest in the South. The forests also contain vast areas of glacial ponds and lakes in their natural setting in the Lake States, several major wild rivers, and at least one "first magnitude" spring. The Appalachian Trail winds through eight national forests on its way from Georgia to Maine.

A final special quality of the eastern national forests is that they are *public* lands, managed according to criteria other than the maximization of private profit.[9] Private landowners, who control the overwhelming majority of forested lands in the East, manage their lands in response to signals given by the market economy. As public lands, the national forests can be managed for public benefits that are not sold in the market economy, such as flood control and wildlife.[10] These are produced by private lands as well, but only through the beneficence of the landowner or as products incidental to other uses. Private owners cannot be relied on to provide these benefits if another use becomes more profitable.

Moreover, the public forests can be managed for a much longer planning horizon than private lands. They therefore can more reasonably be expected to make adequate provision for the demands of future generations and to assume the special role of filling the gaps left by the workings of the market.

We need land in the natural process over a long period of time, as gene pools for special species, for example. Small landowners usually can't plan for long periods, even though they should. The national forests are the only places where this can take place.
DANIEL STILLWELL, Professor of Geography, Appalachian State University Boone, North Carolina

---

Developing the premise that the foremost objective of public lands is to provide benefits for society that cannot be supplied in any other way, The Conservation Foundation believes that,

On the eastern national forests, priority should be given to providing public benefits that cannot be supplied by private land, either because resources are unavailable or an economic incentive is absent.

Implementation of such a policy would have two effects. First, it would remove some of the pressure from the eastern national forests to be all things to all people, making their resources more available to fill those demands for which they are especially suited, including some that have been slighted. Second, removing the national forests from competition with private business would encourage greater use of private forest lands near the national forests, creating new business opportunities for their owners and making improvement of their productivity more profitable.

In view of the special purpose of public lands, there are good reasons for setting goals and establishing styles of management for eastern national forests that are quite different from those that apply to private land. If appropriate opportunities are identified and objectives established for each type of land, their future uses can complement one another in responding to society's demands.

CREATING THE FORESTS OF THE FUTURE

As the eastern United States was settled, then industrialized and urbanized, the original forest that covered it, and that had so impressed the first European colonists, shrank in size and fell drastically in quality. So heavy-handed and heedless of natural values was man's modern occupation of the region that the primeval forest was not just pushed back—a necessary result of human expansion—but virtually disappeared. In its place grew a poor facsimile of the original: a forest, but one lacking in many of the desirable and even impressive qualities that climatic and soil conditions would otherwise permit.

An examination of the current landscape of the eastern United States reveals that only a little over 5 percent of the land is actually devoted to highly intensive (urban, commercial, industrial) use.[11] But these uses are scattered over the landscape in ways that make them seem omnipresent. Despite the fact that more than a third of the total land area of the East and South is forested, there are few places where one can walk for even an hour without encountering roads, fences, signs, powerlines, and other reminders that, in this region, nature has long been subdued by man.

The future is racing toward us. Driving through the South Carolina countryside, you can see trees cut and the land cleared, not just for farms but for more and more industries and shopping centers. Sometimes a developer will bulldoze and even pave over land—then leave it unused. But if we don't save some of the trees and woods and natural areas, our grandchildren won't even know what they're missing. They'll have dirtier air and water and just won't be as healthy as we are.

SALLY BATTLE, League of Women Voters
Columbia, South Carolina

Chapter I describes how the passage of time and protective management by the U.S. Forest Service have returned to a small portion of the East's forest land some of its former character. Indeed, the restoration of environmental quality to the eastern national forests represents one of the great achievements in conservation history. It is all the more remarkable because it has been contemporaneous with what many see as a continuing decline in the health of natural environments elsewhere in the region. But while the decades that have passed since the Weeks Act became law are a long time in the history of institutions, they are not long in the life history of a forest. What the Forest Service has done to date in returning the eastern national forests to productivity is only a beginning.

The Conservation Foundation recommends that in managing its

eastern forests for the long-term benefit of society, the Forest Service give first priority to restoring them to the maximum attainable level of resource quality, emphasizing their potential as natural [12] environments distinct from the man-made environments otherwise dominant in the East. The forest and its products should be used only to the extent that this continuing process of restoration is not interrupted.

This high level of resource quality would represent the best that nature produces, filling in gaps in species, age composition, recreational opportunity, and landscape character which have resulted from human interference with natural processes elsewhere in the East. This is not to suggest an attempt to replicate the wilderness forest that so impressed the first European colonists on their arrival. That was a forest molded not just by climate and evolutionary change, but by insects, disease, and uncontrolled wild fire—forces too dangerous to employ as management tools. They are also crude tools, and wasteful of timber, wildlife, and other resources. Instead, in the future forests, sensitive use of timber management, controlled burning, and other forms of vegetative manipulation should work in tandem with natural processes to restore those aspects of the primeval forest which modern man deems desirable, while still permitting the multiple uses required by law.

These desirable characteristics include high-quality timber stocks, which literally will take generations to produce. Future eastern national forests would boast more climax and "fire-climax" tree communities, with their large, impressive specimens, than would be found elsewhere.[13] Plants and animals which are endangered or extinct in the more settled parts of the region would find refuge on these lands. Expanses of quiet, relatively primitive territory would provide a retreat for recreationists from their normal man-dominated environment. Like the primeval eastern forest, the future eastern national forest would encompass only a moderate degree of landscape diversity; rather, it would itself be a distinctive element in the eastern landscape.

There are two pervasive reasons for preserving such pockets of high-quality forest resources. First, high-quality forests produce particularly desirable yields—including large trees, some forms of wildlife, and certain recreational experiences—that are unobtainable elsewhere. Second, and perhaps more important, the protection of these resources preserves society's options for future uses.

It is not difficult to imagine a future in which some forest products now considered plentiful will be exceedingly difficult to obtain. Early farmers in the Ohio Valley cut and burned countless black walnut trees simply to clear the land. Now, of course, this species is perhaps the scarcest and most valuable of the major American hardwoods. The rapid pace of economic growth and land-use change in recent years makes the possibility of similar errors in resource use more serious than ever.

This is not to suggest that the eastern national forests be managed as though they were national parks; they represent different values and offer a broader range of opportunities for public use and enjoyment. Timber could be cut; gas, oil, and ore extracted; and animals hunted, but only when the judgment had been reached that such uses and activities did not not interfere with the continued recovery of a high level of resource quality and a distinctive natural character—qualities which The Conservation Foundation considers the principal justification for continued public ownership.

The simple fact is, if the Forest Service fails to emphasize these qualities, no one will—and no one else can be expected to. The Forest Service has the land base, both in size and scenic quality. It also has the ability to manage for objectives and horizons not economically feasible for private individuals and corporations.

Its holdings—6 percent of the forested land in the East—stand almost alone as lands that might be returned to something approaching their original high level of resource quality. They present unmatched opportunities for protection and enjoyment of entire environmental systems; for preservation of wildernesses and empty areas; for provision of areas which scientists can compare with the increasingly polluted world about them; for recreational experiences obtainable nowhere else. They are virtually the only large parcels of eastern land left that offer a realistic prospect that these and related natural values can be permanently maintained.

What, then, of the private forest lands in the East? Clearly, regardless of how the eastern national forests are managed, private lands will continue to provide all but a tiny fraction (94 percent in 1970) of the East's total timber supply. They will continue to accommodate intensive recreational development; to supply sand and gravel, oil and ore; to provide forage for animals; and, at times, to provide land for farms, factories, and houses.

Many opportunities exist for increasing both the quantity and quality of what is produced by private forest lands in the East. The 55,000,000 acres of eastern land owned by the forest industry contain some of the nation's most intensively managed timberland, yet even many industrial owners are not applying the most modern management practices. Moreover, these forest owners have little real incentive—beyond public relations and their own feeling of stewardship—to invest their money in managing their land for nontimber uses.

The largest amount of eastern land classified as commercial forest land (268,000,000 acres, or 72 percent) is owned by farmers and "miscellaneous private owners." As a recent report by a panel of professional foresters put it:

> . . . small private nonindustrial forests [have] languished with a minimum of care. Those containing merchantable timber provided a source of emergency cash in time of need. But harvesting usually was "logger's choice," with little attention paid to the need to provide for growing stock for future crops. Even today . . . management of many small forests consists of little more than fencing out livestock and plowing firelines. Some receive no attention at all.[14]

These forests contain nearly three-fourths of the best quality timber-growing lands in the East. Yet, because their owners do not have the knowledge, the inclination, or the financial incentive to do otherwise, lands too often have come to be stocked with trees of poor quality or with species of little commercial value. These lands must be put back into production. Obviously, considerable problems exist in bringing the nonindustrial private lands under better management, but it must be attempted. Every year that improvement of their timber stands is delayed adds another year to the time when the improved stands become merchantable timber.

Those privately owned lands within national forest boundaries have a special role and responsibility. While the owners cannot be expected to manage their land for the same purposes as does the Forest Service, they somehow must be persuaded to manage it in ways that are at least compatible with the uses of the public lands. In some instances, the only way this can be ensured is through acquisition by the Federal Government. In other cases, cooperative planning, local land-use regulation, or federal incentives will suffice. (These alternatives are discussed in Chapters V and VI.)

This broad vision of a future system of eastern forests, with private lands managed more intensively and for a greater variety of uses than at present and with national forest lands emphasizing quality rather than quantity, has implications for each of the major forest uses.

TIMBER

Private lands in the East—both industrial and nonindustrial—should continue to provide the lion's share of the nation's hardwood output and a significantly increased share of its softwoods, including not only pulpwood but also lumber and veneer sizes. Where it is economically profitable, intensive forestry practices should be introduced, bringing a much larger proportion of these lands up to the "state of the art" practiced on the best-managed industrial holdings.

These practices include professional cutting plans, periodic thinning, early removal of cull or inferior trees, restocking with genetically selected varieties, and various techniques for speeding regeneration after harvesting. Other practices, including fertilization, mechanical site preparation, and use of insecticides have merit, yet they are appropriate only when they do not cause erosion, pollute watercourses, or damage soil fertility.

State forest practices laws should ensure that greater use of these intensive forestry practices does not cost society more in environmental damage than it yields in increased timber growth.

The public land in eastern national forests should, in general, continue to specialize in growing high-quality hardwood and softwood sawtimber, with rotations at least as long as those now used. This would not preclude pulpwood management on lands ill-suited for growing sawlogs[15] or for research or demonstration purposes, or the sale of pulpwood from commercial thinnings.

I hope the Forest Service will plan for a slower cutting cycle, in order to provide timber that is fit for furniture, frames, and other permanent human needs, rather than products that are discarded and wasted day to day.

SHERMAN ADAMS, Owner, Loon Mountain Ski Resort
Former Governor, State of New Hampshire
Lincoln, New Hampshire

On the more fertile parts of its southern pine forests, the government should grow sawtimber as economically as possible, while meeting high environmental standards as an example to industry. For instance, since large areas of such forests as the Osceola, Kisatchie, Francis Marion, DeSoto, and Homochitto are flat, there is little danger of erosion. Vegetation returns almost immediately after cutting. Recreational use of these forests consists mainly of hunting, fishing, and driving for pleasure—activities that would be maintained or even enhanced by relatively intensive cutting. Particular care must be taken, however, to protect wetlands and other wildlife habitats in these forests. **Stand conversion and the proper mix of pines and hardwoods are matters to be decided by ecologists and foresters, on the basis of a site's history, soil type, and wildlife-habitat demands.**[16] **Also relevant to this decision is the extent of stand conversion on nearby private lands.** If private owners have converted large areas to pine, the national forest should give more preference to hardwoods. Where conversion is indicated, it should be done by mechanical means, controlled fire, or hand injection of degradable herbicides rather than with aerial application of persistent, potentially dangerous herbicides. And in no case should national forest stand conversion mean, as it often does on private land, trees planted in uniform, farmlike rows.

I think eventually there will be a lot of controversy over the flatland national forests—the forests in the Coastal Plain. They are not as pretty as the mountain forests, but the worst thing that could happen is for them to be turned into tree factories.

LUCY SMETHURST, Chairman of the Board, The Georgia Conservancy
Atlanta, Georgia

Economically profitable forestry may also be practiced on some of the northern hardwood lands in the Lake States national forests, although here the wildlife and recreational constraints often will be tougher. The growing of aspen on very short rotations, although useful as a demonstration project, is a much more suitable practice for private lands than public forests. The public's aspen-growing land should be allowed to go through further stages of succession, returning the original (and also valuable) white pine and longer-lived hardwood cover.

On its more fertile Appalachian hardwood sites, the Forest Service should grow high-quality hardwoods on very long rotations, producing premium wood for the furniture and veneer industries. Even in a well-managed forest, furniture-grade trees are not numerous. High-grading of such hardwoods on private lands, which began at least 200 years ago, is still continuing. The makers of furniture and veneer panels may well be thankful when, years from now, their supplies run out on private lands and they turn to the national forests. **On sites of low fertility or lands where erodable slopes or other constraints make environmentally sound cutting difficult, the Forest Service should not manage for timber at all.**

In the Appalachian forests, cost-effectiveness must often be subordinated to the need to maintain the forest environment. The building of permanent roads should be minimized, unless they would clearly enhance other forest uses. Methods of cutting should be limited to those that eventually will result in a mature, high-quality hardwood forest. These would certainly include both precommercial and commercial thinning, and also would include group and individual tree selection and small clearcuts. Clearcuts—though not necessarily large ones—may indeed be useful in regenerating areas now stocked with inferior trees or in regenerating desirable shade-intolerant species.[17] On the Appalachian hardwood forests, the importance of nontimber uses is such that clearcutting should be used only as a tool for improving forest quality, not as a convenient and economical harvest technique.

The growing of high-quality trees and increased consideration for quality in nontimber uses should make selection-cutting methods more feasible economically than they have been in the past. When the per-acre value of timber rises as a result of better stocking, there will be greater financial incentive to use skyline, balloon logging, and other high-cost techniques.

RECREATION

Public and private lands should complement one another in providing recreation. In general, the public lands should provide the natural resource—the wild river, the hiking trail, the forest wildlife—while nearby private landowners supply the developed facilities, such as highly developed campgrounds.

The public lands in eastern national forests should emphasize dispersed recreation, particularly the type requiring large contiguous parcels of undeveloped land. This includes hiking, backcountry camping, stream fishing, canoeing, and hunting for some species. Developed camping facilities on public land should be simple and rustic, offering no more than the basics needed for visitor health and safety. Perhaps campgrounds should be limited to a few dozen sites, scattered in various parts of a forest to spread out visitors and limit the overuse that tends to occur around the edges of large campgrounds.

There are so many things people can see and do here. The land up here is often said to be fragile, and it is, in the sense that the soil is thin. People must be spread out so as not to overburden and deteriorate scenic places. One of the jobs for the Forest Service is to disperse people so that the land is not overburdened by human use. Some balance must be maintained between the capability of the land to sustain itself and burgeoning public demand.
SHERMAN ADAMS, Owner, Loon Mountain Ski Resort
Former Governor, State of New Hampshire
Lincoln, New Hampshire

Large concentrations of campers around such amenity features as lakes, springs, or streams pose serious social and environmental problems for forest managers. To deal with these problems, innovative campground design, promotion of lesser-known amenity features, and—as a last resort—some form of rationing are preferable to "hardening" the site in ways that detract from the natural feature that attracted people to the area in the first place.

The Forest Service should substantially increase its educational or "interpretive" efforts for forest visitors. The eastern national forests are distinguished more by their diversity of environmental types than by the splendor of their scenery. They are subtle forests, with much of their beauty visible only to those who know what to look for. Increased interpretation would add to visitor pleasure and might also instill a greater respect for the resource. Particular emphasis should be devoted to the history of man in the forest, for man has been interacting with the environment in the eastern forests for many generations, and has left many traces of his occupancy. This is especially true of some of the Appalachian forests, whose human settlement was once intense.

Because statistics show that the largest single recreational use of eastern national forests is driving for pleasure, forest managers should pay close attention to the view from the road. Developed inholdings, even when few in number, often are concentrated along more heavily used forest roads, their signs, food stands, and tourist cottages a visual intrusion into the forest environment. They detract particularly from the concept of the forests as distinctive natural environments, clearly differentiated from the rest of the eastern region. **Upgrading of the visual environment of forest roadsides should be a major goal of future acquisition and cooperative planning.**

In the future, private lands, both currently urbanized communities within the national forests and lands lying outside forest boundaries, should provide most of the highly developed services used by forest visitors. These might include trailer facilities, food stores, rental cabins, swimming pools, riding stables, and ski lodges. As recreational demands grow, these services and facilities can provide new sources of jobs for forest-area residents and profit-making opportunities for local entrepreneurs.

Questions of design control and intensity of development arise from the recommendation that developed facilities ancillary to forest recreation be supplied less frequently by the Forest Service and more frequently by the private sector on private lands. The Forest Service, when it provides or licenses developed facilities, has tended to use rustic designs, minimal signing, and relatively small-scale projects. A new emphasis on private provision of support services should not mean a new deluge of garish souvenir shops, flashing signs, and raucous amusement areas. The new facilities should be forest-related and developed at forest scale. Forest-area farmers, for example, might build a few guest units; other local residents might enter the business of outfitting canoeists or renting horses. The Forest Service could encourage the development of support services related to the forest by including those facilities meeting certain criteria on official recreation maps of each national forest.

WILDLIFE

Wildlife management on the eastern national forests, like timber management, should emphasize quality rather than just quantity.[18] The forests should provide both consumptive and nonconsumptive wildlife "users" with experiences unavailable on private lands.

Wherever feasible, the forests, with their large size and relative absence of development, should provide habitat for solitude-loving wildlife such as bear, cougar, and some of the large predatory birds. Once suitable habitat has been provided and after careful study, wildlife managers may want to try to enlarge the range of these species and, perhaps, to restore animals previously eliminated from large areas. This will not necessarily preclude maintaining sizable populations of more common game animals. Where a choice must be made, however, the public lands should specialize in the more "difficult" species, giving correspondingly less emphasis to those species, such as quail, deer, and grouse, likely to prosper under the short timber rotations that will increasingly be practiced on intensively managed private land. Since some of the scarcer species require the solitude of large tracts of relatively undisturbed forest, their habitat requirements should be considered among the criteria for future acquisitions of land by the Forest Service and for future efforts in cooperative planning for intermingled private lands.

Second, **wildlife management should be directed to habitats and ecosystems rather than single species.** Forest life forms occur not in isolation, but in complex, mutually dependent associations ranging from mammals to insects and from vegetation to soil microorganisms. Recently it has become evident that extinction of the lower forms of life is probably occurring at the same or higher rates as the more obvious extinction of vertebrates. Extinction of lower life forms involves not only loss of genetic material, but also has implications for the higher, more obviously valuable, species which depend on them for food.

Since 1971, national forests in the South have been using a "featured species" system of wildlife managment, which calls for careful study of a species' habitat requirements and the use of harvesting, burning, and other vegetative manipulation to provide suitable habitat. This represents a step forward, yet it focuses on only a handful of animals and birds. This concept should be extended to identify featured ecosystems, with all their interdependent forms of life.[19] Management should be based on the principle that species will thrive if the ecosystem thrives — and a recognition that the converse is not necessarily true.

Finally, **management should give more emphasis — but by no means exclusive emphasis — to nonconsumptive uses of wildlife.** Projections indicate that nonconsumptive activities, such as observation and photography, will be among the fastest-growing categories of recreational demand (page 50). Interpretive programs, trail layout, and wildlife management itself should be more responsive to the needs of nonconsumptive users.

One question sure to arise is the relation of the timber-management program recommended earlier to the amount and composition of forest wildlife. The mature forest, while far from sterile, tends to have a smaller total volume of wildlife than does a forest with many clearings and small trees. A reduction in wildlife volume and perhaps diversity, however, does not mean a reduction in the wildlife values of a forest. Many species, including rare or endangered ones such as the ivory-billed and red-cockaded wood-

A young raccoon peers from a hole in this old maple "den tree" on the Green Mountain National Forest in Vermont.

peckers, require an environment that contains mature or dead trees. These species would be as out of place in a recent clearcut as a deer or grouse would be comfortable. If the intensity of timber management increases on private lands in the East, large blocks of climax forest habitat will become increasingly scarce. In providing such blocks, national forests would reduce their own species diversity, but would raise the diversity of wildlife throughout the region as a whole.

The recommendation that national forest trees be grown on very long rotations would, taken by itself, mean a fairly substantial reduction in these forests' white-tailed deer population. But the impact on hunting opportunity of even a large fall in deer population would be relatively small, for currently the national forests account for only 9 percent of the annual white-tailed deer kill. This impact, moreover, could be mitigated by the clearings left by group selection or patch cuts of mature trees, by the continued cutting of interspersed private land, and by cuts specifically designed for wildlife cover.[20]

In recent years, hunters have watched with concern as huntable lands have been lost through urbanization, posting, and clearing for agriculture. As private lands become less available, hunters have looked more and more to hunting opportunities on public land — including the national forests.

The legitimate demand of hunters for increased wildlife populations should not be met by attempting to maintain on national forests such high levels of wildlife (particularly deer herds) that vegetation or soil is damaged. Instead, the **federal and state governments should encourage increased hunting opportunities on private lands by offering owners financial incentives to improve the productivity of game habitat (by timber management and planting of food crops) and to open their land to the public.** Florida has been a leader in the state leasing of hunting rights on private land; about one-half of the managed public hunting land in Florida is leased private land. This is a particularly appropriate use of national forest inholdings and may be an area in which federal subsidy is warranted, since some of the benefits spill over to the federal lands.

Hunting is a human instinct, more or less—male instinct anyway. I've enjoyed it. I enjoy hunting individually or with my boys. I taught them to hunt, and I've been hunting since I was 12 years old. My family has grown up hunting. Last year my wife and daughter each shot a doe. I hunt bear, deer, and turkey. We use the game that we kill. I cut up the meat, and wrap and freeze it, and the family helps.

Most hunting skills involve patience, knowledge of the woods, topography, knowledge of game habits and game trails, and knowledge of firearms. Being a hunter, I spend quite a bit of time in the woods, and you

have to be observant. The kind of berries and fruit tell you where to hunt, the time of year it is tells you when to hunt, and the number of people in the woods tell you how to hunt—whether to "stand," or to "drive."

The game is wary and not altogether defenseless. They have better senses than you or I and the forest is their home. Just as I would be hard to find in this house, they would be hard to find in their woods if they wanted to hide. It's only when they make a mistake that you're able to see them.

I think they have too many roads in the national forest. I know some of them are necessary to get the timber out, but I think they should be spaced further apart. And, after the timber is taken out, the roads should be closed.

The more roads they put in, the more crowding there is in the forest. A solitary-type hunter finds it harder and harder to enjoy his type of sport. I think that's the main difference between hunters who like the roads and hunters who don't like the roads. A lot of hunters use jeeps and snowmobiles to find game, and then get off and shoot it. I own a jeep, but I don't use it to hunt—but that's their way of hunting.

JACK SKINNER, Service Center Foreman, Bell Telephone Company
Warren, Pennsylvania

WILDERNESS

In passing the 1974 Eastern Wilderness Act, Congress reaffirmed its intent to set aside not just virgin areas, but also ones in which vestiges of past human activity are fading away. Even under this broad view, the number of places of sufficient size and remoteness to qualify is not large. **This rarity makes it urgent to give statutory protection to those areas that are of wilderness quality, yet remain undesignated.** Many of these have been identified by national and local conservation groups. It may be useful to cite three examples of these lands, along with capsule descriptions of the kinds of resource values that designation would protect:[21]

> Big Island Lake (6,000 acres, Hiawatha National Forest, Michigan) — Ponds, lakes, swamps, and mixed conifer and hardwood forests are the principal features on this glacial moraine topography. Here and there stand impressive groves of beech. If the traveler goes quietly, by foot or canoe, his chances of seeing coyotes, raccoons and other mammals — perhaps even a rare wolf or moose — will be enhanced.

> Wild River (20,000 acres, White Mountain National Forest, New Hampshire) — possesses an unusual diversity of flora and fauna that makes it scientifically attractive. On the marshes and grasslands live moose and beaver. In remote woods of birch, spruce, and fir are bears and bobcats, plus other predators such as coyotes.

Cheoah Bald (19,000 acres, Nantahala National Forest, North Carolina) — wide open summit [with] good views of the misty southern highlands... covers all sides of the mountain as well as its 5,062-foot summit. On one side is the scenic Nantahala Gorge. There are cliffs, cascades, deep virgin forests on steep slopes, spires of blue slate, and hiking trails that penetrate the remote and quiet coves.

Because suitable areas are in such short supply, it is likely that even additional designations will provide fewer areas than wilderness recreationists call for. To meet this demand, **The Conservation Foundation recommends that, in addition to designated wilderness — and not as a substitute for it — a number of areas in eastern national forests be managed as "low management intensity" areas.** In these areas, there would be restrictions on permanent roads and structures of any kind and on intensity of timber management. Timber cutting would be allowed, but only with methods not requiring permanent roads. Entry by any sort of vehicle would be infrequent. Cuts would generally be limited to final harvest of mature trees; indeed, the logical candidates for such areas are those in which there is no economic payoff to thinning or other forms of active timber management.[22]

The private sector cannot be expected to provide wilderness, except perhaps accidentally or through private conservancy organizations. The relatively few suitable areas on eastern public lands are, quite simply, all there is. Management policies must recognize their uniqueness and protect them.

We shouldn't be sidetracked by different interpretations of the term "wilderness." The fact that foresters tend to define wilderness as something close to a truly virgin condition should not distract us from the fact that, in general, their objectives and those of wilderness users are very similar. If the public's demand for wilderness can be met with long-rotation hardwood management, then the Forest Service must gain the confidence of the public by committing large acreages of eastern national forests to this type of management. Such a strategy satisfies the multiple-use mandate and still preserves future management options. And preserving options rather than foreclosing them should be the primary objective in national forest planning.

ALBERT F. IKE, Associate Director
Institute of Community and Area Development, University of Georgia
Athens, Georgia

As a general proposition, the national forests in New England are under management which is fairly conservative in terms of timber production. Because of this, the New England national forests do not pose the kinds of problems for water managers that you find in some other areas, such as California, where the cutting of certain redwood stands results in the deposit of huge amounts of sediment in the streams. Because of this, water planners in the east do not have a sense of urgency about dealing with forest problems. This is partly a reflection of the fact that water management is pretty effectively considered by the national forest staffs. Now if they were doing a lousy job, we'd be worried about it.

R. FRANK GREGG, Chairman, New England River Basins Commission
Boston, Massachusetts

WATERSHED

Water runoff is an unmarketed, often unplanned product of both public and private lands. In most parts of the East, good watershed management means that cutover and other disturbed sites are promptly revegetated, that roads are properly built and maintained, and that the wastes of man and animals are retained on site. Society has come increasingly to insist that landowners, both public and private, assume responsibility for any adverse external effects the use of their land may generate.[23] Thus **prompt revegetation should be required by state forest practices laws. Point and nonpoint sources of water pollution should be regulated by vigorous enforcement of existing state and federal pollution-control laws. The Forest Service should help states develop such regulations for forest land and should set an example for other landowners by strictly adhering to these standards in its own land-use practices.**

RANGE AND GRAZING

Use of eastern forest lands by livestock should be encouraged, but only if it can be done in a manner that is not damaging to the forest or its soils. In practice, this probably means permitting grazing only on pine forests and mature hardwood forests where young timber cannot be damaged. Private lands usually are better suited for this purpose than are the national forests, for they do not have the constraints imposed by multiple use. This is particularly true with respect to wildlife, which should be given first preference in the use of the forage produced on public lands. Nevertheless, the Forest Service can provide a useful service by carefully researching the potential for greater use of the eastern forest range, particularly in the South and the Lake States. Grazing use of national forest land, where appropriate, will probably serve mainly as an example to private owners, although there may well be some areas of public lands on which grazing can be practiced without interfering with other uses.

A WORD ABOUT MINING

Minerals, the mining companies are fond of saying, are where you find them — and it should come as no surprise that many are found in inconvenient locations. Some deposits may be under actual or potential wilderness areas; others lie under prime timber-growing land; still others are associated with streams and aquifers. This capriciousness of nature is indeed an argument for preserving considerable flexibility in federal minerals policy. It is no excuse, however, for the present system of exploration and leasing of federally owned minerals, a system that is hopelessly inefficient and outdated. Nor does it excuse passive acceptance of the results of historical circumstance which finds so much of the eastern national forest subsurface in private hands.

The federally owned minerals in eastern national forests should not be arbitrarily withdrawn from development. But it is tempting to suggest this, unless the way in which federally owned minerals are exploited is drastically improved. This would involve a number of changes in law and practice. First, **the existing morass of federal mining laws should be replaced by a single law providing for discretionary leasing of all types of minerals, on all types of federal land, by competitive bid.** There would be no preferential right established by private prospecting, and a single federal leasing agency would be required to weigh the national need for the mineral in question against the damage to the environment and to surface use. The government would have full discretion as to whether to grant a lease and under what conditions.

Second, **the present prospecting and "discovery" system should be replaced by a systematic government inventory of mineral deposits on federal lands, most appropriately conducted by the U.S. Geological Survey.** Better information about mineral deposits would help the government secure maximum revenues for the resources that it owns. More important, it would enable better planning for meeting future mineral demands.

Finally, **when the government's holdings of nonrenewable resources are known, they should be exploited most conservatively.**

As in the case of timber, the private sector is likely to give far more weight to the demands of the current market than to those of succeeding generations. This leads private mineral owners to exploit their holdings at relatively rapid rates over fairly short periods of time. This may make economic sense for the private mining company, but has much less justification on the public lands. **Where their immediate explotation would interfere with other uses of the land, minerals in public ownership should be conserved for future use, or to cushion the impacts of future national resource emergencies.**[24] Such a policy, to be sure, involves an economic loss — the postponement, perhaps indefinitely, of the revenue that could be raised by selling these minerals. Yet in a world in which future values, opportunities, and problems are quite unclear, taking the cautious course has its attractions.

ADDITIONAL OPPORTUNITIES ON THE EASTERN FORESTS

The Forest Service should legitimately feel pride in its restoration of the eastern national forests to their present condition. Now, 66 years after passage of the Weeks Act, **The Conservation Foundation recommends that the Forest Service assume two additional major, long-term tasks: the restoration of lands devastated by strip mining and the establishment of a major jobs program on the national forests.** Both are consistent with the historical mission of the agency, yet represent a considerable new commitment of energy and resources.

The millions of acres of abandoned strip-mine sites in a sense are similar to the eastern national forests at the turn of the century — lands depleted for quick profits, and then abandoned. In their present state, they not only are unproductive, but create recurring erosion and acid drainage. Most are in the Appalachians or the Ohio Valley; a quarter of a million acres scar the land near the Wayne National Forest alone.

Restoring these lands to productivity — particularly those lying within and adjacent to existing eastern national forests — will be difficult, time-consuming, and expensive. But this is a task the Forest Service could appropriately undertake, since the bulk of this land would be most easily restored as forest land. Trees would cover the scars and slowly begin to restore the soil. The recovery of landscape quality would add to the scenic and other values of the nearby federal forest land. And the potential funding source, a tax on newly strip-mined coal, would be logically levied at the federal level.

The eastern national forests can also serve as a source of new jobs. Since 1946, Congress has set a goal of national full employment. Nevertheless, the economy frequently has been far off the mark. Currently, there seems to be considerable political interest in direct federal job provision through the creation of new public-service jobs. Lawmakers interested in such programs would do well to look back to the amazing success of the Civilian Conservation Corps (1933-42), both in providing employment and in accomplishing useful work. Many of the campgrounds, trails, and shelters built by the CCC in national forests remain; some are sadly in need of maintenance or reconstruction. The restoration of timber quality on eastern forests can be traced back, in part, to the planting of millions of seedlings by the CCC. Any new federal program of job creation should include as a major element the provision of jobs associated with the eastern national forests. Such an effort could build on the Forest Service experience with such successful, though small, programs as the Youth Conservation Corps, the Job Corps, and the Older Americans Act employment program.

I think we need another CCC. We need it as a permanent but flexible way of dealing with unemployment. You're sitting here in Brunswick, Vermont, about 600 yards from the site of a 1938 CCC camp. It's right up the road. Kids from Rhode Island, Roxbury, New Bedford, Fall River—their names are still carved on the boards of the spring house. One of my colleagues got his start in forestry right here in 1930—he helped build the camp.

BRENDAN WHITTAKER, Chief, Information and Education
Agency of Environmental Conservation
Montpelier, Vermont

The Forest Service appears eager to accept this responsibility, claiming recently that it could provide as many as 200,000 Youth Conservation Corps positions across the nation, as well as 120,000 jobs under other manpower programs.[25] Such jobs have a number of attractive features. They are located in rural areas, many of which suffer considerable unemployment of the "structural" type — that is, intractable to the usual solutions of monetary and fiscal policy. Many forest jobs require relatively unskilled labor and involve low capital costs per worker. Finally, the workers employed

> The Forest Service should provide employment opportunities in forest-related jobs to minority people who live in urban areas. Minorities just don't realize that there are jobs out there in the public sector—that the Federal Government employs thousands of people in forestry and land-management jobs.
>
> Many minority people are trained in recreational services—people management, not land or resource management. Both the Forest Service and the Park Service have numerous opportunities to deal with youths, and urban park managers are particularly experienced in people-handling. This is a skill which the Forest Service should welcome.
>
> BETTE WOODY, Boston University
> Former Massachusetts Commissioner of Environmental Management
> Boston, Massachusetts

would be net additions to the working labor force, not substitutes for persons now employed by the private sector. A greatly expanded jobs program not only would make an immediate contribution to social welfare, but also would be a considerable help in continuing the restoration of resource quality to the eastern national forests.

Creating Incentives

To assert that a piece of land is most suitable for a given use or combination of uses does not ensure that the land will be so used; that requires the creation of incentives or penalities. For private landowners, the incentives are potential profits; for public land managers, the incentives are political support and adequate funding. Thus, to shift uses between public and private forest lands requires the creation of an appropriate incentive structure. Three types of incentives should help implement the forest objectives recommended by The Conservation Foundation: a fee structure for national forest recreational use that demonstrates the value of recreational benefits to society, while recapturing some of the public costs necessary to produce them; incentives to increase the profitability of timber management, grazing, and recreation on private forest lands; and a system to allow local governments to profit from forest uses other than timber production and mining.

CHARGING FOR RECREATION BENEFITS

In terms of budget appropriations, recreation has been a stepchild of national forest management. Between 1963 and 1972, the final Forest Service appropriation for timber-sale administration was 99 percent of the level planned for by the agency.[26] For fire protection, the level was 70 percent of the request; for range management, 79 percent. Recreation, however, was allocated only 44 percent of the amount requested. Undoubtedly, a major reason for these deep slashes in funding is the knowledge in the Office of Management and Budget and elsewhere that, while expenditures for management and development of the physical resource base produce current or future revenues, recreational expenditures return little or nothing to the federal treasury.

Under the current system of very low or no charges for recreational use of the national forests, this is certainly an accurate assumption. Between 1973 and 1975, receipts from timber sold from eastern national forests averaged $33 million annually; revenues from recreational use fees were $1.5 million. It is little wonder, then, that expenditures on timber management are viewed as an investment, while those on recreation are considered an expense.[27]

This disparity in revenue generated obscures the fact that the value to recreationists of the 44,000,000 visitor days spent in the eastern national forests is probably far greater than the value of the timber that these same forests produce. In fact, if an arbitrarily low value of 75 cents were assigned to each visitor day, total recreation benefits would be as high as total timber revenues.

Given the tremendous social value of eastern national forest recreation, it is unfortunate that it produces such a minor impact on what legislators and budget officials perceive as the forests' "profit statement." **The Conservation Foundation therefore recommends that modest fees be charged for recreational use of the eastern national forests — and indeed all federal lands — not only to generate useful revenue, but to affirm to decision makers the intensity of the public demand for high-quality forest recreation.**[28]

In its 1970 report, the federal Public Land Law Review Commission recommended that:

> A general recreation land use fee, collected through sale of annual permits, should be required of all public land recreation users and, where feasible, additional fees should be charged for use of facilities constructed at federal expense.[29]

Clearly, there are problems connected with the imposition of entrance fees to public land, particularly to such large and fragmented tracts of land as the national forests. The testimony heard by Congress after a brief and inconclusive federal experiment with an entrance-fee system suggests that it is feasible only if it applies to a wide range of federal lands (not simply the national forests) and if it relies primarily on a low-cost annual permit, much like a fishing license.[30] With such a system, enforcement would be limited to spot checks, and would entail minimal costs and little interference with the enjoyment of users.

> *If forests are to be used for recreation, we should pay for what we use. People should be aware that what they are using is costing somebody something. We may also have to restrict recreational use of the forest—by limiting the number of people in a given area at any one time.*
>
> DANIEL STILLWELL
> Professor of Geography
> Appalachian State University
> Boone North Carolina

Even if a uniform fee system is not adopted, two other modifications to the present system of fee charges on eastern national forests could significantly increase recreation revenues. First, **fees for developed campgrounds, now set at very low levels, should be raised to equal the amount charged for similar facilities on private campgrounds.** In most cases, this would involve increasing fees to between $3.00 and $6.00 per night, the range of fees charged by the private sector. These fees should be collected only when the revenue raised promises to be significantly higher than the government's cost of collection. This probably would mean that fees would be levied only in summer and on weekends, times of the highest use and the highest degree of crowding. Fees for other developed facilities, such as boat-launching ramps, also should be charged at market levels. Appropriate fees also might be collected at specially developed areas for motorized sports vehicles.

Second, **charges should be made for recreational use of wilderness areas and for boating on designated wild and scenic rivers.** Wilderness recreationists make rather heavy demands on forest land. Not only do they require that uses of the land incompatible with wilderness be foregone, but the very nature of their enjoyment means that only a low ratio of visitor days per acre can be accommodated. Wilderness recreation — clearly a reasonable use of some national forest land — should help pay its own way. At the very least, it should reimburse the government for the revenues foregone when land is reserved for this use.

Wilderness-use charges need not be prohibitively high. The average net (1973-75) revenue from timber cutting per acre of eastern national timber land is estimated at no more than 95 cents (Chapter II, page 28). The Forest Service reports: "Informed judgments by wilderness managers of different areas with varying use pressures suggest [about one-half of a visitor day per acre per year] may be close to a desirable upper limit on some wildernesses."[31] At this limit, a wilderness-use fee of less than $2.00 per visitor day would more than offset the losses of revenue to the government from foregoing timber cutting.

Wilderness-use fees probably should be specific to a particular wilderness area. First-time or casual users would pay a daily or weekly permit fee at the nearest ranger station.[32] Along with their permit, they would receive information about trails, fire restrictions, and other material important to their safety and enjoyment. Selective use of trail advice could be used to disperse groups fairly evenly over the wilderness and, at times when crowding threatened, the number of permits issued could be limited. Frequent users, who would have less need for orientation, could purchase an annual permit allowing unlimited entry to the area.

Proposals for fees for forest recreation raise obvious questions of equity. But statistical profiles indicate that national forest recreationists tend to have average or higher incomes (Chapter II, page 71). The fees proposed are modest relative to what forest recreationists willingly spend for equipment and recreation travel. And it is difficult to argue convincingly that forest recreation is as deserving of public subsidy as are education, health care, and income security — with which forest-recreation expenditure requests must compete when the annual national budget is developed. It is little wonder that, each year, recreation budgets suffer.

Forest recreationists often have supported reasonable use fees and are likely to continue to do so if assured of high quality in return for the charge. Those who balk at this recommendation might reflect on the long experience with hunting and fishing licenses, in which user fees have created a powerful professional constituency that protects the interests of the user groups. Healthy forests, with impressive trees and a variety of wildlife, with clean water and an absence of development, are worth the modest price it would take to obtain them.

INCENTIVES FOR PRIVATE PROVISION OF FOREST SERVICES

Three major reasons help explain why owners of the 268,000,000 acres of private, nonindustrial forest land in the East have done so little to improve their timber management: Owners lack the necessary knowledge; the land is held for nonforest purposes; or — most important — intensive forestry simply has been unprofitable.

The Forest Service, through its State and Private Forestry program, has made considerable efforts to educate private forest owners about suitable methods of growth and harvest. This effort should be continued, but in much closer coordination with the activities of the National Forest System. Methods of timber culture employed experimentally on national forests should be widely publicized among forest owners in the region. Moreover, special services should be provided to owners of land near or within national forest boundaries, to encourage them to practice environmentally sound forestry.

Many owners of private, nonindustrial forest land in the East hold the land for reasons other than timber production.[33] Some may use it for recreation; others speculate on rising land prices; still others have inherited the property and have no definite plans for its use. If these forest owners are to be persuaded to manage more intensively, they must be convinced that timber management is profitable and can be accomplished in ways that do not preclude these other uses. Many owners, disturbed by scenes of clearcuts they have observed or read about, appear to think that timber harvest is necessarily followed by long-term aesthetic damage, soil damage, and erosion. **The national forests, with their mandate of multiple uses and a sustained yield of forest products,** should serve as models for private forest owners who seek to manage for similar purposes in a spirit of land stewardship. To demonstrate techniques and economic feasibility to small private landowners, managers of the eastern national forests should experiment with uneven-age management systems and with methods of timber culture suitable to small-scale management units. Special programs should be developed to transfer this knowledge to the small landowner. Small portions of the forest might be set aside as demonstration plots, with tours for landowners.

Over the years, the national forests have perfected and demonstrated many techniques of timber management that later were adopted profitably by large timber companies, particularly in the South. Now it is time to focus attention on the small landowner, recognizing that he, like the Forest Service itself, uses his land for multiple purposes.

Undoubtedly, the greatest barrier to more intensive management of eastern private forests has been the low profitability. As Minckler points out, "There is no way a small private woodland owner can justify the present high price of forested land and other fixed costs on the basis of timber values alone."[34] Individuals will continue to hold forest land for a variety of reasons. Policies should be developed not so much to ensure that trees are grown on these lands as to see that the trees grown are merchantable ones and that environmental quality is not unreasonably degraded. Thus, policy recommendations with the goal of increased timber production should be aimed less at reducing the fixed costs of holding timberland (as by preferential tax assessment) and more at reducing the marginal costs of good timber management or increasing the return.

There are at least three routes toward this goal, each of which deserves careful study. One is direct federal sharing of the costs of reforestation and stand improvement,[35] which has been practiced on a limited scale for some time under the Agricultural Conservation Program and, more recently, the Forestry Incentives Program. The Forest Service reports that, in 1974, "about 90 percent of all forest practices on nonindustrial private land was cost-shared under ACP or FIP."[36] A second approach would allow landowners to deduct the cost of forest improvements from their annual taxable income, including nonforest income. Currently, such expenditures must be capitalized and are deducted from timber income as trees are harvested.

The final — and most direct — incentive is to allow the price of timber to rise to levels that make it profitable to raise trees as a crop. It has been obvious for decades that timber prices have remained relatively low only because the huge stock of trees that constituted the original American forest was being harvested without the costs of replacement and the depletion of supplies being factored into market prices. For most of our history, these old trees have been cut on private lands, but since World War II, an increasing proportion has come from western national forests.

Only recently, as private old-growth stocks have been depleted and national forest supplies subject to environmental constraints, have timber prices increased enough to justify intensive forestry in the private sector. Despite the attraction of low-priced timber, it is important to realize that low prices discourage long-term private investments in timber growing. **Liquidation of old growth timber in national forests should not be hastened or increased in the cause of maintaining artificially low timber prices.**[37] **Rather, the goal should be to achieve prices that reflect the cost of replacing what is cut — prices high enough to call forth from private landowners the requisite future supply.**

Private landowners also need incentives to provide recreational support facilities and to open their land to hunting and other forms of recreation. Increased charges for national forest use should make it more profitable for private landowners to provide campgrounds and similar facilities. Public leases of hunting land or trail easements also should provide revenue to landowners.

INCENTIVES FOR LOCAL GOVERNMENTS

Local governments tend to view the national forests in terms of the local revenues they generate. Until recently, local governments were paid a fixed proportion of the particular forest's annual receipts. Among its other faults, that system gave local governments a financial incentive to push for intensive timber cutting and mining on the public lands — those forest activities that produce the most revenues — in preference to other kinds of uses. This bias was corrected in legislation enacted in 1976. (See pages 225-231.) Under the new system, local governments should be able to give greater weight to other areas in which their financial interest is interconnected with the use of the national forests — the expenditures of forest visitors, the goods and services bought locally by the Forest Service, the local employment generated, the provision of watershed, and public recreation.

The Forest Service should commission a modest research program to identify and quantify the net benefits that communities derive from their proximity to a national forest. This would include not just direct and indirect revenue from commodity uses, such as logging and mining, but that generated by recreation, hunting, fishing, and other noncommodity uses. If, as seems likely, the benefits of the forest to local governments are considerably greater than the direct payment they now receive from forest revenues, this information might be persuasive to local officials. If, on the other hand, cases are found in which the forest does little to help the local economy, some management changes may be in order.

Most of the local residents seem happy that they have something in the forests that attracts people and brings in additional income. Communities within the national forests are deeply affected by changes in forest use. As recreation increases, communities will have to adjust—change from logging to managing resorts, for example. It won't happen overnight, but it will come.

DANIEL STILLWELL, Professor of Geography, Appalachian State University
Boone, North Carolina

On balance, The Conservation Foundation's suggested goals for the eastern national forests should result in increased economic activity in nearby communities and increased revenues to local governments. Greater investment in timber-stand improvement, additional investment for noncommodity uses, and rehabilitation of strip-mined land would all require additional workers, most of whom should be drawn from local communities. Greater use of selection cutting rather than clearcutting would require more timber markers, as well as an increase in the labor intensity of the cutting process itself. New recreational opportunities associated with an improving forest resource would provide additional local jobs in the private sector. Selective Forest Service acquisition of forest inholdings would free local governments of the costs of servicing small scattered developments.

Relations between the powerful Federal Government and small, often impoverished local governments are inherently unequal.

Recognizing this, the Forest Service has historically moved with restraint, even when it believed that it was clearly in the right. In the future, as federal and local authorities become increasingly involved in joint projects, particularly in the area of cooperative planning, the Forest Service will have to devise new methods to achieve its goals by incentives and cooperative action.

The Civilian Conservation Corps at work in the 1930's near the nation's first CCC camp—Camp Roosevelt on Massanutten Mountain on Virginia's George Washington National Forest.

REFERENCES

Chapter III

1. As Krutilla and Fisher put it: "There is no known technology for the production of a new natural environment, which is the accident of geomorphology, weathering, and biological processes involving a time span far exceeding human planning horizons." John Krutilla and Anthony Fisher, *The Economics of Natural Environments* (Baltimore: Johns Hopkins University Press, 1975), p. 12.
2. If all land within the purchase boundaries is considered regardless of ownership, 46 of the forests are larger than 250,000 acres.
3. Missouri's Mark Twain National Forest, newly created by merging two existing forests, has 1,400,000 acres, but they are scattered in blocks over a rather large area.
4. Four of these holdings are larger than 1,000,000 acres each: International Paper Company tracts in Maine and Arkansas, Great Northern Paper in Maine, and Buckeye Cellulose Corporation in Florida (data provided by the American Forest Institute).
5. Other large tracts of public land, but less than 250,000 acres, are Baxter State Park (Maine); Wharton State Forest in the New Jersey Pine Barrens; Catskill Forest Preserve (New York); Land Between the Lakes National Recreation Area (Kentucky-Tennessee); Okefenokee National Wildlife Refuge (Georgia); the proposed Big Cypress National Fresh Water Preserve in Florida; and some state and county forests in Wisconsin.
6. Lester Klimm, "The Empty Areas of the Northeastern United States," *Geographical Review*, Vol. 44, No. 3 (July 1954), pp. 324-45.
7. Dennis Durden, "Use of Empty Areas," in F. Fraser Darling and John P. Milton, *Future Environments of North America* (Garden City, N.Y.: Natural History Press, 1966), p. 479.
8. There is no easy way to estimate the rate at which "empty lands" have been disappearing. However, when Durden (1964) surveyed 20 of the empty areas of Connecticut that Klimm had identified in 1954, he found that only 11 still qualified. All but one of these remaining open spaces were on public land (Durden, p. 492). See also Florida State University, *Florida Reference Atlas*, "Map of Uninhabited Areas of Florida."
9. National forests make up about two-fifths of the total public forest land in the eastern states. One-fifth is in state forests, while the rest is set aside for parks or wildlife refuges or is used by the military.
10. The national forests' lack of response to "economic" constraints and incentives has been much criticized. The criticism is fair to the extent that it takes the Forest Service to task for producing economic goods (principally timber) in uneconomic ways. For example, it has been charged that revenues from some timber sales have not even covered the costs of managing the sale and of building roads needed to take the wood out of the forest. Such criticisms simply amount to saying that when the national forests are managed to produce the same goods as private business, they should respond to the same incentives.
11. U.S. Department of Agriculture, Economic Research Service, *Major Uses of Land in the United States* (1969), Agricultural Economic Report No. 247 (Washington: U.S.D.A. 1973), p. 37.

12. Natural environments are defined here as those whose dominant features are either the result of natural processes or of human interventions which mimic natural processes. The future forest would not be "natural" in the sense that all natural forces would be allowed to run their course. Neither, however, would it be artificial, in the sense that an alfalfa field is an artificial replacement for a natural meadow, or a pine plantation replaces mixed-age, randomly distributed pine-hardwood forest.
13. A climax forest is a biotic community of vegetation that represents the equilibrium adaptation to the area's soil and climate. A "fire climax" is a subclimax community held back from further succession by periodic forest fires, or by other, man-made disturbances having the same effect. Most of the southern pine forests are fire-climax communities.
14. Kenneth B. Pomeroy and John Muench, *The Challenge of Private Woodlands* (Washington: American Forestry Association, 1973), p. 5.
15. A good example of such an area is the sand-pine ecosystem that covers much of Florida's Ocala National Forest.
16. The question of what is "natural" is a difficult one, for wild fire, now largely controlled, has been the formative element in the southern pine ecosystem.
17. See Leon Minckler, *Woodland Ecology* (Syracuse, N.Y.: Syracuse University Press, 1975), pp. 70-71.
18. Technically, national forest wildlife is the property of the individual states, which manage it through fish and game commissions. These commissions set hunting seasons and stock streams. The Forest Service has the important responsibility of providing the habitat. Under the 1974 Sikes Act, the states and the Forest Service have been engaged in cooperative planning to relate Forest Service habitat programs to state wildlife management programs.
19. For example, one participant at the Atlanta regional conference conducted by The Conservation Foundation remarked on the casual destruction of the wiregrass ecosystem through mechanical soil preparation in the regeneration of southern pine.
20. Patch cuts have more "edge" per acre than large clearcuts. Moreover, according to the Forest Service, "Because they remain accessible to deer, heavy selective cuttings appear to have an important advantage over clearcuttings." U.S. Forest Service, *Report of the Chief, 1970-71* (Washington: GPO, 1972), p. 25.
21. Descriptions are from Ann and Myron Sutton, *Wilderness Areas of North America* (New York: Funk and Wagnalls, 1974).
22. See Marty's study of the return to management intensification, which showed that returns on investment in silvicultural treatments during the life of a stand vary widely with the fertility of the site. Robert J. Marty, "Economic Effectiveness of Silvicultural Investments for Softwood Timber Production," U.S. President's Panel on Timber and Environment *Report* (Washington: GPO, 1973), pp. 141-47.
23. For example, in Just v. Marinette Co. 201 N.W. 2d 761 (1972), a federal court prevented a landowner from filling a wetland because of the potential flood impact on downstream properties.
24. The setting aside of federal petroleum reserves in California many years ago now makes it possible for them to make a modest addition to domestic oil production.
25. U.S. Forest Service, *A Recommended Renewable Resource Program* (Washington: U.S.F.S., 1975), pp. 511-12.

26. "Status of Financing a Development Program for the National Forests" in Daniel R. Barney, *The Last Stand* (New York: Grossman, 1974), Appendix 5-3, pp. 157-58.
27. The argument was made earlier, however, that some timber "investments" may well be unprofitable for the Forest Service.
28. The National Park Service, which charges a $2-per-car entrance fee at Everglades National Park and Shenandoah National Park, raised just over $1 million from these units alone in 1975.
29. U.S. Public Land Law Review Commission, *One Third of the Nation's Land* (Washington: GPO, 1970), p. 203. The earlier Outdoor Recreation Resources Review Commission had recommended, "Public agencies should adopt a system of user fees designed to recapture at least a significant portion of the operation and maintenance costs of providing outdoor recreation activities that involve the exclusive use of a facility, or require special facilities."
30. U.S. Congress, House Committee on Interior and Insular Affairs, *Proposed Amendments to Land and Water Conservation Fund Act*, 92nd Congress, 1st Session (Washington: GPO, 1971).
31. U.S. Forest Service, *The Nation's Renewable Resources: An Assessment*, p. 101. It is likely that many eastern wilderness areas could accommodate greater than average intensity of use.
32. It was noted above that permits for wilderness use probably will need to be required simply to protect the resource.
33. See surveys of owner attitudes cited in Pomeroy and Muench, *The Challenge of Private Woodlands*, pp. 15-17.
34. Leon Minckler, *Woodland Ecology*, p. 37.
35. In the East, provision might usefully be made for sharing the cost of wildlife improvement cuts, with the proviso that limited public hunting be allowed.
36. U.S. Forest Service, *A Recommended Renewable Resource Program*, p. 297.
37. The Forest Service says its current harvest scheduling is based on maintaining an even flow of timber by "extending the conversion process for the old growth over a long period." *Program for the National Renewable Resources*, p. 307. This principle is hinted at, but not explicitly mentioned, in the 1976 National Forest Management Act.

CHAPTER IV

forest legislation and the forest service

Establishing goals for the eastern national forests is only a first step toward ensuring their future ability to meet the changing demands of society. The laws that govern management of the forests and the structure and traditions of the Forest Service can affect attainment of these goals.

LEGISLATION

A 1974 Forest Service compilation of "principal" laws governing its activities lists 44 separate pieces of legislation. Broadly speaking, the laws authorize financial and technical assistance to state and local agencies and private landowners, provide for forestry research, and guide the administration of the National Forest System.

While the compilation includes the National Environmental Policy Act, which increasingly influences Forest Service activities, it does not include other federal environmental legislation, such as the water- and air-pollution control acts, which also constrain Forest Service actions. Nor is the annual appropriations act mentioned, although funding from Congress largely determines what the Forest Service does. "The Appropriations Act gives us our marching orders," said one Forest Service official.

Five fundamental laws in particular affect management of the eastern national forests: the Organic Administration Act, the Weeks Law, the Multiple-Use Sustained-Yield Act, the Forest and Rangeland Renewable Resources Act, and the National Forest Management Act.

119

The *1897 Organic Administration Act* sets forth basic purposes and management principles for the national forests (then called forest reserves). National forests were to be established "to improve and protect the forest within the [national forest] boundaries or for the purpose of securing favorable conditions of water flow, and to furnish a continuous supply of timber for the use and necessities of citizens of the United States."[1]

Before passage of the Organic Act, no legislation set forth the legal uses of the forest reserves. This act defined the ways in which the forest resources could be used by those who had settled in and around the western forest reserves. Thus settlers were permitted to cross national forest land to reach their land, to build wagon roads as necessary, and to use two acres for a schoolhouse site and one acre for a church site. Prospectors could enter the forests to search for minerals. Streams and lakes within the reserves could be used for domestic water supply, mining, milling, and irrigation.

The Organic Act also contains important provisions governing timber harvesting. It specifies the nature of the timber that can be removed from the forest ("dead, matured, or large growth trees . . . as may be compatible with the utilization of the forests thereon . . ."), and requires that each tree "before being sold shall be marked and designated." These Organic Act provisions were the basis for the successful 1973 challenge by environmentalists to Forest Service timber-harvesting practices, particularly clearcutting, in the Monongahela National Forest. The court found that the Forest Service was cutting immature trees that were not properly "marked and designated," in violation of the Organic Act. That decision and subsequent ones in Alaska and Texas stimulated congressional enactment of the 1976 National Forest Management Act, which repealed the Organic Act's limits on timber removal and established new guidelines for timber harvesting on the national forests.

The Weeks Law, enacted in 1911, can be considered the organic act of the eastern national forests, providing legislative authority for federal acquisition of forest land.[2] As originally enacted, it authorized the acquisition of "forested, cut-over or denuded lands within the watersheds of navigable streams . . . necessary to the regulation of the flow of navigable streams. . . ." A provision of the 1924 Clarke-McNary Act added "for the production of timber" to the original Weeks Law language as a reason for national forest purchase. Other provisions of the Weeks Law affecting national forest purchase deal with state approval of land to be acquired and the sharing of national forest revenue with states for the support of schools and roads. As an incentive to the states, the Weeks Law also provides for federal assistance in forest-fire protection. Significantly, the Weeks Law contains no findings, statement of purpose (other than preserving the navigability of streams), nor congressional policy toward the national forests to be established. In the context of the times, it was apparently considered sufficient that the forest land—forested, cut-over, or denuded—be acquired, and the Forest Service surely would know what to do with it.

The Multiple-Use Sustained-Yield Act of 1960 stands as the basic statutory guide to national forest management.[3] It also raises profound questions. The act requires that the national forests be administered for (and the law lists them in alphabetical order) "outdoor recreation, range, timber, watershed and wildlife and fish purposes." Under sustained yield, the renewable resources of the forest are to be managed to achieve "in perpetuity . . . a high level annual or periodic output . . . without impairment of the productivity of the land."

But with competing demands for the forest resources, most attention has focused on multiple use. The language of the act defines multiple use as

> . . . (T)he management of all the various renewable surface resources of the national forests so that they are utilized in the combination that will best meet the needs of the American people; making the most judicious use of the land for some or all of these resources or related services over areas large enough to provide sufficient latitude for periodic adjustments in use to conform to changing needs and conditions; that some lands will be used for less than all the resources; and harmonious and coordinated management of the various resources, each with the other, without impairment of the productivity of the land, with consideration being given to the relative values of the various resources and not necessarily the combination of uses that will give the greatest dollar return or the greatest unit output.

Multiple use is a dogma of forest management and the linchpin of national forest land-use planning. Shortly after passage of the Multiple-Use Sustained-Yield Act, then-Forest Service Chief Richard E. McArdle explained the significance of the legislation:

The legislative history of this act directs that in making application of the principle of multiple use to a specific area, equal consideration is to be given to all of the various renewable resource uses, but this does not mean using every acre for all of the various uses. Some areas will be managed for less than all uses, but multiple use management requires that there be two or more uses.[4]

The public interest in the national forests can vary from one period to the next, depending on which group has control of the media and popular sentiment. It has been a most unfortunate situation that over the past 10 or 15 years we have seen a polarization between preservation on one side and utilization (or conservation, as I prefer to call it) on the other. I don't think these groups are very far apart. We both basically want the same thing: we want a clean, healthy environment, we want to grow trees, we want more trees growing than are being removed. As an industry, we are very much committed to the sustained yield principle. Within the eastern national forests, there is room at the present time to accommodate all the various pressures and demands which are being made upon them. You can't set a priority for any particular user group; all have a legitimate right to use the forest.

Most uses are compatible, although maybe not at the same time on the same area. For wildlife, for water yield, you need some timber harvesting. For recreation you need roads. Who builds the roads? The timber industry. You could develop a recreation area and leave it for 10 or 15 years and then move the facilities somewhere else and permit the area to be harvested. In this manner, recreation and timber harvesting can be compatible.

PETER R. MOUNT, Chairman, Western North Carolina Forestry Commission
Leicester, North Carolina

The Society of American Foresters has defined multiple use concisely as "a strategy of deliberate land management for two or more uses which utilizes without impairment the capabilities of the land to meet different demands simultaneously."[5]

If the Organic Act was considered too precise in its strictures with regard to timber harvesting, the Multiple-Use Sustained-Yield Act is so broadly drawn as to be susceptible to almost any interpretation. While the Forest Service views the Multiple-Use Sustained-Yield Act as statutory confirmation of national forest management policy dating back to Pinchot, others believe that it is espoused by the Forest Service because it permits the widest possible management latitude. Writes one Forest Service critic, "Whenever the agency as a whole senses an external threat to its considerable autonomy, it shouts 'foul!' and points to possible violations of the Multiple-Use Act."[6] The late Orris Herfindahl said that multiple use, while "certainly applicable in many cases, [sometimes] turns out to be just a slogan serving to camouflage the complete sacrifice of one use to others."[7] And Marion Clawson has questioned its economic efficiency: "In practice, multiple use has all too often meant a little of everything everywhere, including timber management on uneconomic sites."[8] Yet this act may also be too limiting, particularly in its failure to acknowledge the aesthetic and environmental benefits that flow from forest lands to broad regions. Recently, the West Virginia Forest Management Practices Commission suggested that an "aesthetic use" be added to the multiple-use list. And the Forest Service's plan for the Allegheny National Forest acknowledges citizen proposals that "environmental amenities become one of the emphasized multiple-use items."[9]

I see an evolution in the policy of the Forest Service. They are going to have to examine the priorities of multiple use. These inevitably will change toward recreation. The Forest Service cannot be only timber managers. In some areas, where recreation demand is high, the forests may be managed much like national parks; in other forests, timber will have priority. The multiple-use concept is good, but we have to have local selection of priorities. So many of our problems result from use of terms people don't really understand—like clearcutting. It's a red herring. Many people don't understand that certain species need sunlight to reproduce, sunlight that comes through removal of trees. Likewise, many people don't realize that recreation has an economic value. If a person doesn't understand the economic value of recreation, but only knows that timber brings in dollars, then it is impossible to talk about priorities.

DANIEL STILLWELL, Professor of Geography, Appalachian State University Boone, North Carolina

---

Much national forest land, even that not specifically designated for wilderness or recreation, has been assigned a use emphasis, often tacitly. On the timber-poor Massanutten Mountain unit of the George Washington National Forest in Virginia, recreation appears to be the dominant use; in the timber-rich coastal plains forests of the South, timber clearly is the de facto dominant use of large forest areas. The forest plan for the White Mountain National Forest lists four "management area" types, each accommodating several uses, but with the use emphasis of each ranging from dispersed recreation to timber.[10] Most forest plans do not zone for

> Multiple use, as I understand it, does not require every management unit to be managed for all "products" at equal intensities or with equal priorities. The nature of any given site may easily justify its designation for a primary use (recreation, water, timber production), with other uses subordinate to that primary use. There may be a few areas where you have the opportunity to manage for all five uses, but that's rare. Then there may be areas where you are restricted to only one use, but that's rare too.
> ALBERT F. IKE, Associate Director, Institute of Community and Area Development, University of Georgia Athens, Georgia

uses in the way proposed for the White Mountain, though many forests prescribe special management for distinctive scenic areas or along well-traveled roads.

New to the statute books are two landmark laws which should equal the Multiple-Use Sustained-Yield Act in importance for future forest management. The 1974 *Forest and Rangeland Renewable Resources Planning Act* (RPA) establishes a process for determining natural resource management objectives and a program for managing the resources of the national forests. The planning process required under RPA is intended to be national in scope and long range, and should integrate private-sector capabilities and performance with those of the public lands. It is to result in a presidential policy, implemented through congressional appropriations, to achieve national renewable resource goals. RPA makes national forest planning a part of the RPA program, requiring that the Forest Service "develop, maintain and, as appropriate, revise, land and resource management plans for units of the National Forest System..."[11]

The link between RPA and the Forest Service's land-management planning was reinforced with passage of the *National Forest Management Act of 1976* (P.L.94-588) in the waning days of the 94th Congress. Designed to deal with the legal impasse over timber harvesting on the national forests brought about by court decisions in the Monongahela suit and in Alaska (see page 120), the new act repeals the Organic Act's strictures on timber harvesting and makes national forest plans the means for regulating harvest practices. The act also substantially amends many provisions of RPA, in an effort to prevent environmental abuses resulting from the harvesting of timber and to ensure equal consideration for, and protection of, *all* the forests' renewable resources. Timber is not to be favored over other forest resources. The amendments add new requirements both for the direction and content of the RPA program and national forest plans.

Future RPA programs are required to evaluate multiple-use relationships among the various renewable resources of the forest and establish national goals that "recognize the interrelationships between and interdependence within the renewable resources." The program is required to "recognize the fundamental need to protect, and where appropriate, improve" air, water, and soil quality. The private landowner also is to be informed of opportunities to improve his land.

New requirements for national forest planning form the heart of the National Forest Management Act. If NFMA works as intended, the system-wide regulations it requires will provide guidelines to encourage resource management sensitive to the characteristics of individual forests via the forest plans. A process to regulate timber harvesting has been grafted onto national forest planning. The Forest Service is required to develop and issue regulations to control timber harvesting and particularly prevent the misuse of clearcutting in national forests. Guidelines are to "provide for a diversity of plant and animal species," an effort to deter conversion of mixed hardwood forests to exclusive pine, and to permit timber harvesting only where it will not cause damage to watersheds and where streams can be adequately protected. Clearcutting is to be permitted only where "it is determined to be the optimum method;" the clearcuts are to be in patches, shaped and blended to the natural terrain, with maximum limits established for size.

The NFMA is another milestone in the congressional construction of national forest policy. But while the act deals comprehensively with the management of the forest resources under multiple use, it does not address a multiplicity of broader questions, including the fundamental role of the forests themselves both in national and regional contexts, and their relationship to privately owned land and resources.

A NEW STATUTE FOR THE EASTERN NATIONAL FORESTS

Disappointingly, this most recent legislation affecting national forest management does no more than earlier legislation to emphasize distinctions in management needs for the eastern and western forests. It can be argued that in the course of resources planning under RPA, administrative action to implement the National Forest Management Act, and the Forest Service's own planning process, a distinct management approach will emerge for the eastern national forests. This may be the case, but judgment must be based on present evidence. Currently, no Forest Service document forcefully calls attention to the distinctive eastern national forest features—their fragmented ownership pattern, proportion of the regional resource base, and proximity to populous metropolitan areas—or points to any special management direction for the eastern forests as a whole. Eastern national forest supervisors are left to resolve forest use conflicts with no clearer overall objective than multiple use, with all its ambiguities.

The eastern national forests should receive statutory attention as a separate subsystem of the National Forest System, in recognition of their distinctive qualities and the special opportunities they offer. Special regional attention has precedent in national forest legislation. The original Wilderness Act had its greatest applicability in the West. Additional wilderness legislation was written in recognition of the special characteristics of the healing areas of the East that were deemed not of sufficient quality under the "pure" wilderness definition of the original Wilderness Act. While the authority of the Weeks Law has been used to acquire some western land, it was designed to be the legislative vehicle for establishing national forests in the East.

Eastern national forest legislation should include the following in its statement of findings:

1. The national forests of the East constitute the largest single category of public land east of the 100th meridian and incorporate most of the wilder and undeveloped areas of substantial size and scale remaining in the East.

2. The eastern national forests can be expected to be used for a variety of legitimate purposes, but accumulated demands, intensified by a growing population with more leisure time, mobility, and requirements for forest commodities, have placed intense pressure upon these forests.

3. The eastern national forests present unique management opportunities for long-term social benefits. Among the benefits are recreation; valuable timber species; wilderness; habitat for game and nongame wildlife and fish, including rare and endangered species; and high-quality water for use far from the forests.

4. The forests' locations near major eastern metropolitan areas offer opportunities for activities and the provision of forest products easily accessible to urban populations.

5. The pattern of federal forest ownership in the East fosters dynamic interrelationships between the public and private lands, creating both problems of use and opportunities for the environmental and economic enhancement of these regions.

Other legislative provisions should establish general objectives for the eastern forests: "It is the policy of the Congress that the national forests of the East be managed with the objective of providing renewable resources of superior quality, of species and variety such as the private sector is unwilling or unable to provide; further, the balancing of uses under the concept of multiple-use sustained-yield shall take account of the unique or outstanding values of each forest or forest unit."

This would establish a direction for forest management consistent with the objectives set forth in Chapter III.

A new eastern national forest statute should also establish the protection and enhancement of natural environmental values as a major objective for the eastern forests. Sometimes termed "aesthetics" or "amenities," the natural environment concept is much broader than beauty or attractiveness. It represents a whole new category of national forest uses and values, responsive to the new types of forest users who have emerged in recent years. Some of these—in general younger and more urban-oriented than previous users, with more leisure time and greater mobility—are articulate and influential advocates of dispersed recreation opportunities, such as backpacking, rock climbing, caving, and whitewater canoeing. They are joined by a much larger group of users who enjoy the forests for a wide range of nonconsumptive uses—outdoor photography, wildlife study, plant and flower identification, and simply walking. Together, these groups constitute a strong force for maximizing the unique natural opportunities offered by these forests.

The best description of this concept is set forth in the forest plan for the Allegheny National Forest, which defines amenities as those things in an environment which:

- make it pleasing in appearance,
- lend a sense of well being, both economically and socially,
- contribute a sense of lack of tension,
- provide opportunity for solitude and reflection,
- contribute to the preservation of plant and animal life, and maintain the variety of both,
- demonstrate tender, loving care of the land itself.[12]

While The Conservation Foundation favors the statutory route to eastern national forest management direction, administrative courses could usefully be pursued as well. When the administration of the then-forest reserves was transferred to the Bureau of Forestry in 1905, Secretary of Agriculture James Wilson sent a letter to Gifford Pinchot, Chief of the Bureau of Forestry, which spelled out the spirit as well as the details for forest reserve management. Could the Secretary of Agriculture not issue a comparable policy directive on eastern national forest management? While not as binding as a statutory mandate, a policy statement from the Secretary of Agriculture, to be prominently displayed in eastern national forest manuals and plans, could prove an effective vehicle for insuring administrative attention to the distinguishing features and circumstances of the eastern forests.

OTHER STATUTORY CHANGES

Although this report is concerned with the forests of the East, some recommendations for improvement clearly have system-wide applicability. Particularly, several statutory changes would benefit not only the eastern forests, but the entire system. There is, of course, danger in exposing long-established statutes such as the Organic Act and the Multiple-Use Sustained-Yield Act to legislative tampering. But it seems evident that amendments to the two laws can and should provide policy direction more appropriate to present and future forest objectives and uses.

Nothing in the present Forest Service lexicon formally recognizes the enjoyment of natural environmental values as a distinct and positive use. It can be argued that the five uses now recognized in the multiple-use law—timber, range, outdoor recreation, watershed protection, and fish and wildlife—are broad enough to encompass such interests. But inferring new uses from those now accepted, such as outdoor recreation and watershed protection, only adds to already serious problems of definition. More importantly, the failure to recognize natural environmental values as a separate, identifiable category of beneficial use makes it difficult to plan, acquire, and manage national forest lands specifically for this purpose.

Consequently, **The Conservation Foundation recommends that a sixth use—protection and enhancement of natural amenity and environmental values—be added to the Multiple-Use Sustained-Yield Act.** This would provide a clear focus for planning and management directed to these purposes and would also permit a more accurate assessment of the costs and benefits at stake in adopting alternative management strategies.

In a clause often overlooked by those who focus on the timber and watershed purposes for national forest establishment, the Organic Act provides that national forests are to be established to "improve and protect" the forests. This principle is given conditioned reaffirmation in the National Forest Management Act, which requires the federal Renewable Resources Program to "recognize the fundamental need to protect, and where appropriate, improve the quality of soil, water, and air resources."

However, submerged as this is in the act's directive for the preparation of the Renewable Resources Program, it lacks the prominence that would come from a position in the Organic Act.

The "improve and protect" clause should be considered the overriding principle of national forest management. To this end, the Organic Act should be amended to give deserved prominence to the objective of forest improvement, protection, and restoration, and to encompass all the uses of the forest—including environmental enhancement—in the list of purposes for forest establishment. The amendment might read: "National forests are established to improve, protect, and restore the forests for present and future generations; to secure favorable conditions of water flow, furnish a continuous supply of timber, provide outdoor recreation, wildlife and fish habitat, and forage; and to protect and enhance the quality of the environment."

The Organic Act should also be amended to include a clear statement of mission for the Forest Service. While legislation establishing some other federal agencies has included such a statement —a philosophical expression of duty that serves as a guiding precept of policy and activity—the Forest Service has had none. Evolving as it did from a small bureau of forestry to the present large and diffused organization, it was left with the nuts and bolts directives of the statutes, but no broad institutional charge.

A provision of the National Forest Management Act attempts to correct this deficiency:

> The Forest Service, by virtue of its statutory authority for management of the National Forest System, research and cooperative programs, and its role as an agency in the Department of Agriculture, has both a responsibility and opportunity to be a leader in assuring that the nation maintains a natural resource posture that will meet the requirements of our people in perpetuity . . .

Yet that provision stops short of forging an institutional mission linked to the management of the national forests.

By contrast, the mission of the National Park Service is clear, and specifically related to the purposes of the parks:

> . . . to conserve the scenery and the natural and historic objects and the wildlife therein and to provide for the enjoyment of the same in such manner and by such means as will leave them unimpaired for the enjoyment of future generations.[13]

A statutory definition of an agency's responsibility certainly does not end all the uncertainties or the political controversy over individual management decisions. It does, however, provide a constant charge and challenge to the agency, and, one might argue, constrain tendencies toward temporary bureaucratic aberrations.

In interviews, a number of Forest Service staff members—district rangers, forest supervisors, and others—were asked what they considered their "mission." Often the initial response was in terms of multiple use and sustained yield. But upon reflection, all expressed a far deeper sense of continuing responsibility for the basic well-being of the land and resources under their management. This commitment is not adequately reflected in statutory language. To correct this deficiency, an amendment to the Organic Act might read: "It shall be the responsibility of the Forest Service to conserve the resources of the national forests, providing for their use in such ways as are consistent with the fundamental capability of the land and other resources; and to protect and improve the forest for the benefit of present and succeeding generations."

Such a statement makes it clear that the Forest Service's role is stewardship and that it is responsible for ensuring the availability of forest resources of high quality, not only for the present generation but for future ones as well. This is a charge consistent with the present Forest Service role, accommodating both use and protection, transcending place, time, and the political bickering that punctuates much national forest policy making.

NATIONAL FOREST PLANNING

To translate the mandates of national legislation into actions on individual forests, the Forest Service has developed a complex and sophisticated planning process. While the intent is sound, the planning documents issued to date are disappointing in their failure to provide distinctive regional direction for the special opportunities and problems of the eastern national forests.

The Forest Service's land-management planning process began evolving in the late 1960's. Intended to be comprehensive, detailed, and continuous, it has as one of its major goals to make national forest management consistent with forest resource capabilities and national, regional, and local economic and social objectives. The process requires that a comprehensive 10-year management plan be developed for each forest, along with more detailed plans for in-the-field management of smaller units in each forest. If these plans function as intended and are effectively integrated with the RPA program (see page 136), the land-management plans will translate national policy into field-level management decisions.

The new forest planning process attempts to shift away from traditional functional planning. Under the old system, each plan focused on a single aspect of forest management, such as timber or fire protection or recreation. Although each functional plan became a component of a forest multiple-use plan, it usually was simply incorporated between covers along with the others, with little effort to assess the impact of the management plan for one resource upon other resources. The new land-management process, however, strives to construct an integrated program for all the resources, based upon the productive capacity and natural characteristics of the land itself. Ideally, the forest plan should guide forest managers "in allocating the use of National Forest System resources, helping to balance the needs of people with the capabilities of the land."[14]

LAND MANAGEMENT PLANNING IN THE EAST

Eastern national forest management planning is guided by *System for Managing the National Forests in the East*, a slim manual issued in 1970.[15] Though now outdated because of refinements in the planning process (the *System*, for example, makes no reference to individual forest plans, which are now the principal planning documents), it provided the initial procedural direction for a new process. The *System* divides the United States east of the 100th meridian into seven planning areas (see map, page 133), based on similarities in geography and forest characteristics. The Appalachian Mountains from the Allegheny National Forest in Pennsylvania to the Talladega National Forest in Alabama constitute a single planning area, for example. A guide, providing direction for all the forests within the region, was subsequently developed for each planning area.

But anyone looking in either the *System* or the guides for distinctive treatment of the eastern forests based on characteristic factors, such as fragmented land holdings or the proportion of the area's resource base in national forests, will be disappointed. None of these documents offers direction for managing the eastern forests in general, or the forests within the planning areas, in any way significantly different from forests in any other area. For example, the Appalachian guide establishes these management priorities:

> Protect and enhance the air, water, soil, and natural beauty including the unique nonreplaceable natural, historic, and archeological features.
>
> Create a variety of quality outdoor experiences for rural and urban dwellers with emphasis on simple, restful, uncrowded association with nature.
>
> Achieve a high standard of management on one of the finest, most extensive hardwood forest ecosystems in the world.[16]

Only the third refers to a distinguishing feature of the eastern forests; the other goals would be equally applicable to the vast, consolidated forests in the west.

The System should be replaced by a new document which both sets forth specific steps in the planning process (as the current System does) and establishes special program directions for the eastern forests. This direction should be carried through in the seven regional guides, with adjustments that recognize special regional forest characteristics and user needs. The replacement to the *System* should provide a discussion of issues, problems, and opportunities peculiar to the East (especially the fragmented landholdings, proximity to metropolitan areas, acquisition policy, relationships with state and local governments and private landowners, and opportunities for cooperative action) and set forth management objectives tailored to the eastern forests. Revised area guides

PLANNING REGIONS FOR EASTERN NATIONAL FORESTS

should describe the resource situation of each region and establish area forest management direction. They should also provide a framework for integrating planning for the national forests with other regional planning efforts, such as Section 208 Areawide Waste Treatment Management Plans under the Federal Water Pollution Control Act, and river basin planning under the Water Resources Planning Act.

INDIVIDUAL FOREST PLANS

At the forest level, national direction is converted to plans for field-level management. The forest land-use plan, says the Forest Service Manual, "is the key forest land-use coordinating tool. It is at the national forest level that servicewide and planning area objectives, targets, and direction are interfaced with specifics about the land."[17] Forest plans establish an overall framework for the detailed plans for each of the smaller management units—usually a discrete ecological system, such as a watershed or mountain range. The format for a forest plan is prescribed in the Forest Service Manual in a way that suggests these plans are to be quite comprehensive. Each is to deal with 10 specific elements, including a description of the major laws governing national forest management, a statement of national and regional objectives, an analysis of management alternatives for a given forest and their probable consequences, and a description of the program selected for guiding management of the forest for the next decade.

Each forest plan also is to describe the "management situation" in a statement which "specifically relates the Forest Service planning objectives to a finite land area."[18] This section is to provide data on the region's population and local economic, social, and cultural needs, and information on how the national forest fits into the regional setting. "The management activities of other landowners on related lands should be spelled out together with the special management implications" for the national forests. "Opportunities for cooperation [should be] identified . . ." In another section, the forest plan is to assess the capability of the land, "expressed in terms of both measurable resource items and less easily quantified amenity values present." These guidelines should provide forest planners with the opportunity to identify the special values of the individual forest, its place in the region, and implications for future management direction. Unfortunately, this opportunity goes unrecognized in most forest plans.

The loose-leafed forest plans, laced with maps and statistical tables, generally follow the Manual's direction. There is a brief summary of the national legislation under which the Forest Service operates, a recapitulation of the management direction promulgated in the area guide, a history of the forest, an assessment of its current situation, sometimes a discussion of current issues affecting the individual forest, and proposals for management direction (with alternatives) interwoven with an analysis of the potential environmental impacts of different management courses. Not surprisingly, within the specifications of the Manual, some forest plans are quite comprehensive, while others are superficial.

But a commonality of approach pervades almost all current forest plans. It would be unfair and inaccurate to say that Forest Service planning policies require cookie-cutter conformity; the outline in the Manual permits variations among areas to be taken into account. A few forest plans, such as that for the White Mountain, recognize the unique attributes of specific forest areas and attempt to shape management to accommodate them. But the overall thrust in these documents is toward uniformity in forest planning, not innovation. Planners clearly feel that departures from the standard are not encouraged and must be justified. Balancing uses within each forest, rather than recognizing interforest differences in resources and demands, appears to be the common objective. Under these circumstances, plans sensitive to the particular distinctions of individual forests are rare. **Forest plans should be better attuned to the special characteristics of each forest—such as scenic features, forest topography, opportunities for special kinds of recreation, superior timber productivity, and patterns of land ownership that affect forest use.**

Despite instructions in the Manual that adjacent lands and potential impacts from private land uses upon the forest resources be analyzed, forest plans generally ignore adjacent private land, even when its use has major implications for national forest management. Private land, if mentioned at all, usually is considered only in terms of its potential for acquisition. This neglect is exemplified in the Ocala National Forest plan's treatment of a privately owned area of high natural value known as Salt Springs, a wedge of land undergoing development within the forest. "Although contained within the legal boundary of the Ocala National Forest, this 10,000-acre unit, managed for its recreational features, is privately owned," the plan states. "Therefore, no further mention of it is made here."[19] **Forest plans should describe the interrelationship of public and private land and should assess the potential of nearby private land to meet public needs, a critical variable in determining the best use of the public's forest land.**

Most forest plans fail to describe and analyze how the forest fits into the environmental, social, and economic framework of its region. Population and economic trends are discussed shallowly, if at all. **Forest plans should analyze the condition of the local and regional environment and the present and future role of the forest in regional environmental enhancement.** Among the social and economic issues that should receive attention are local life-styles, trends in leisure-time activities by metropolitan users of the forest, local employment trends, and job-creating opportunities on the forest. In some areas, much of the work already has been done by state planning offices or regional planning councils, and could easily be incorporated into forest plans.

Finally, forest plans should evaluate the effectiveness of the state, regional, and local governments and agencies which control private land and resource use. Inadequate management of adjacent private land interferes with the Forest Service planning goal of efficient management of the public's forest resources. If a local jurisdiction has no zoning ordinance and ineffective subdivision controls, the forest plan should say so—and explain what this means to the public's land, water, trees, wildlife, and recreational opportunities. Additionally, forest plans should explain how forest management fits into—and, ideally, reinforces—regional planning efforts such as river basin plans, areawide wastewater management plans, and coastal zone planning. Inconsistencies between forest plans and regional plans should be pointed out.

A distinctive management approach has yet to emerge in planning for the eastern forests. The situation is not static, of course, for the Forest Service must continually redefine its planning process and adjust it to the new requirements of the Renewable Resources Planning Act and National Forest Management Act.

THE RENEWABLE RESOURCES PLANNING ACT (RPA)

RPA grew out of a realization—stimulated by resource shortages and interminable controversy over timber supply—that no one knew the true extent of the national supply of forest resources. Even in those areas where an effort had been made to collect data, the information and subsequent decisions often were not weighed against impacts on other renewable resources. Then there was the continuing charge that, despite the multiple-use mandate, the Forest Service favored timber over the other forest resources. In the Congress, no long-term objective governed funding for resources management. Commented Senator Hubert Humphrey, "We work too much on an ad hoc basis in the Congress and executive branch, moving from crisis to crisis—applying policy and funding Band-Aids and aspirin to long-range problems that require permanent treatment. This is particularly true in our forests and rangelands." [20]

In response to this problem, RPA was developed to provide an information base for future renewable resource policy and programs, to require the formulation of long-range policies and programs, and to ensure that budget requests are consistent with national policy. The RPA Program, according to the Forest Service, "provides the basis for a total management process that will guide and facilitate future long-term policy making, decision making, budgeting, and on-the-ground activity."[21] The RPA process is to look four decades into the future.

Forest resources are renewable, yes, but only over a long period of time. They are not renewable in terms of the average person's lifetime. People do not think in terms of a couple of hundred years, but in terms of their own lifespan, and to them, a cutover forest is not renewable. It's an emotional, psychological problem that leaves out logic—on both sides.
LUCY SMETHURST, Chairman of the Board, The Georgia Conservancy
Atlanta, Georgia

Briefly, RPA requires the Forest Service to conduct a periodic assessment of the nation's renewable resources, both those in government ownership and those privately owned. (After an initial submission scheduled for 1975, the Assessment is to be updated in 1979, and at 10-year intervals thereafter.) Following submission of the Renewable Resources Assessment, the Secretary of Agriculture is to transmit to the President a 40-year Renewable Resources Program for management of the National Forest System, for research, and for cooperative programs with State and Private Forestry, to be updated every five years. The President is to forward it to Congress, accompanied by a statement of policy, which can be rejected or revised by Congress. This policy statement is to serve as the basis for future administration budget requests which are to state "in qualitative and quantitative terms" how they meet the objectives of the policy document. The Secretary of Agriculture is to issue an annual RPA progress report to Congress and the public.

RPA provides that national forest land-use and resource-management plans be "coordinated with the land and resource planning process of state and local governments" as well as other federal agencies. Further, information from the Assessment and Program may be used by state and local governments in their own planning. The Assessment identifies how Forest Service programs relate to other "public and private activities" and discusses "important policy considerations, laws, regulations, and other factors expected to influence and affect significantly the use, ownership, and management of forest, range and other associated lands." Thus, RPA links national forest lands and private lands.

The Forest Service strongly supported enactment of the Renewable Resources Planning Act as landmark legislation. In the small-type charts and graphs and bureaucratic prose of their first Assessment and Program, Forest Service officials see the embryonic protoplasm of a progressive and comprehensive national resources program. However, this vision is far from realized in the first Assessment and Program, submitted to Congress in March 1976. Completed in just 16 months, the Assessment (345 pages) and Program (more than 1,100 pages, including agency and public questions and comments) represent a monumental effort. For all their bulk, deficiencies exist. The Assessment is strong in areas where there had been past statistical work—particularly in timber—and weak in resource areas where there had been less statistical collection and analysis. Information is sparse on private-sector production. And the very size of the Program hinders a layman's assimilation of the material.

The principal achievement of the Program is the development of at least a rudimentary structure for assessing the costs and benefits of resource-investment options. It identifies six renewable resource systems: land and water, timber, outdoor recreation (and wilderness), wildlife and fish habitat, range, and human and community development. Alternative goals are arrayed for each system—usually different levels of activity or investment higher or lower than present uses. Then, the various options for each system are combined with others and integrated into eight alternative total programs, with differing levels of effort and activity emphasis.

With numerical increases of both costs and outputs, the final national management direction selected from the alternatives in the 1976 Program can best be described as using more intensive—and expensive—management to give something to everyone. It calls for an increased supply of recreation, with an emphasis on types of recreation that take advantage of natural features and do not require intensive development; a moderate increase in wilderness; a substantial increase in wildlife habitat; the provision of forage, with costs commensurate with benefits while protecting the productivity of the land; increases in timber supply and quality "in an environmentally sound manner to the point where benefits are commensurate with costs;" conformance with minimum air- and water-quality standards while improving soil productivity and air and water quality; and increased human and community development activities.

The President's policy statement is to be the capstone of the

Sheep graze in the crest zone of the Mount Rogers National Recreation Area in the Thomas Jefferson National Forest in Virginia.

administrative processes established by RPA. Unfortunately, President Ford's statement accompanying the first Assessment and Program stressed programmatic uncertainties related to decisions on federal tax policy and budget priorities, rather than policy resolution. While restating the basic goals of the RPA document, the President omitted any mention of his funding intentions.

Dr. Marion Clawson has identified analytical deficiencies in the Assessment and Program, and also pointed out the difficulty a lay reader encounters in comprehending the massive technical documents.[22] Particularly disappointing is the fact that neither document provides the information base, discussion of issues, or program direction necessary for effectively dealing with special regional characteristics, especially those of the eastern forests. The Assessment offers little data for any region on resource capability and public needs. Nor is there a substantive discussion of critical issues on a regional basis, despite a requirement of the act that the Assessment include "a discussion of important policy considerations expected to influence and affect significantly the use, ownership, and management of forest, range, and other associated lands."

Congress required that the RPA process be national in scope. Yet Section 3 of the act provides latitude for the development of program elements for separate forest regions: "The program transmitted to the President may include alternatives, and shall provide in appropriate detail for protection, management, and development of the National Forest System . . ." But the Program does not go "into appropriate detail" as far as regional needs, problems, and opportunities are concerned. And the reader who wants to know what the Assessment and Program mean for the forest he uses will search the documents fruitlessly.

There are many issues of distinct regional importance. For example, harvesting old-growth timber, while it has national timber-supply implications, affects only western forests. For the East, numerous regional issues merit discussion in the Program. Most of these relate to the role of the eastern forests as "minority" forest landholdings and the management questions this raises: What are future acquisition goals and priorities in the East? How is the Forest Service to deal with conflicting uses of private land adjacent to the forests? How are questions of use allocation to be resolved in those forests where landholdings are especially small and fragmented? What should be the posture of the national forest manager in dealing with local officials and private landowners? Are new directions needed for Forest Service assistance to state and private forestry in the East? If so, what are they?

Such questions cannot be answered with hard statistics, of course, but they deserve discussion as critically important issues in the development of a resource management program. Especially in light of RPA's mandate for coordination of forest plans with those of state and local governments, the lack of attention to the Forest Service's relationships with its neighbors cannot be viewed as insignificant.

The Program is important, too, as the basis for future presidential budget requests. Several budget items are especially important to the eastern forests, particularly appropriations for land acquisition under the Weeks Law. For this purpose the absence of substantive policy discussion of eastern forest acquisition objectives is especially disappointing.

THE FOREST SERVICE

Neither the national forest laws nor the planning process can be understood apart from the structure and characteristics of the Forest Service, the agency responsible for managing the national forests. The organization of the Forest Service, the makeup of its staff and their professional attitudes, and certain personnel policies can affect such substantive issues as the role assigned to timber production in the eastern national forests, national forest acquisition, and cooperative action among the Forest Service, states, and local units of government.

The formulation of national forest policy and its implementation can be pictured as a complex circuit through which information is transmitted from the district ranger, through the Forest Service hierarchy, to upper echelons of the administration and, finally, to the Congress. Policy, budget, and program directives flow back through the circuit. The transformer in the circuit is the Forest Service, responsible for policy, program, and budget proposals, and, after processing by the administration and Congress, for implementation.

A CHANGING ROLE

In his classic study *The Forest Ranger,* Herbert Kaufman wrote:

> The Forest Service was born late, grew fast, has a broad range of functions and a large and finely divided organization. Leading an agency under these conditions, especially in the face of strong and carefully marshalled opposition from politically powerful interests, is not an easy task; the leaders have many things to watch, and respond to, simultaneously.[23]

In its early years, when its primary assignment was the custodianship of a threatened resource, the Forest Service was subjected to bitter criticism and legislative assault from those whose exploitative, and frequently illegal, use of the public's resources was thwarted by dedicated rangers. Feelings ran so high that there were serious efforts to dismantle the Forest Service. Legislation was proposed to restrain its activities; salary increases were refused by Congress. Forest Service morale plummeted.[24]

Until World War II, forest management emphasized protection, not necessarily because the Forest Service wanted it that way, but because there were few demands on the national forests. As one commentator has observed, "For almost five decades, until the housing boom at the end of World War II, the major forestry activities on the national forests were protective, competitive interest in purchasing timber was limited, and the volume sold was a small part of the national production."[25]

But since World War II, the Forest Service's role has changed from custodian to manager. The postwar housing boom and the development of new uses for fiber prompted an apparently insatiable national demand for timber. With private land failing to match demand, attention turned to the national forests, including

I started working in the forest industry in the 1920's and negotiated many timber sales contracts with the Forest Service. Back then, the Forest Service was a caretaker rather than a silviculturist. It didn't do anything spectacular except improve its fire-fighting capabilities. Congress has always been niggardly with providing funds for timber-stand improvement which is unfortunate, since it is a project that ultimately pays for itself.

SHERMAN ADAMS, Owner, Loon Mountain Ski Resort
Former Governor, State of New Hampshire
Lincoln, New Hampshire

---

the eastern units which now had merchantable stands of timber. During this postwar period, the forests also began to attract recreationists in ever-increasing numbers.

Faced with competing demands for a limited land base, the Forest Service was required to manage the land to satisfy as many of these interests as possible. And in this era of management, the Forest Service finds itself assaulted from all sides. A former chief of the Forest Service is reputed to have commented, "I am supported by the pressures that surround me."[26]

The Forest Service fulfills responsibilities defined and assigned to it by federal statutes, further modified by its annual appropriation. Decisions determining the uses to which the forests are to be put—ranging from the preparation of the budget for submission to Congress, to timber sales, construction of campgrounds, and placement of roads and trails—are made by Forest Service officials in Washington, regional foresters, supervisors, and district rangers.

ORGANIZATION

The Forest Service is a rigidly hierarchical organization. Michael Frome writes: "The Forest Service is in no way a military organization, but like an army with many field units, it has a single central authority that operates through a decentralized organization to ensure application of uniform principles."[27] Some of its characteristics are reminiscent of an efficient military force: the uniform which most Forest Service personnel wear in the field, esprit de corps, camaraderie, and professionalism. An official from another federal agency refers to the Forest Service as "the Marine Corps of the bureaucracy."

At the peak of the hierarchy stands the chief of the Forest Service. Deputy chiefs, associate deputy chiefs, staff officers, and a host of other aides in the Washington office serve as his staff, counseling the chief and helping to transmit directives throughout the organization. The line of authority flows from the chief to key ad-

---

The Forest Service has some pretty able people. As a government agency, I always thought it was the best—morale is high, and it knows what its mission is. It's dealing with a wonderful resource—land and timber.

BRENDAN WHITTAKER, Chief, Information and Education
Agency of Environmental Conservation
Montpelier, Vermont

---

ministrative officers at succeeding levels of the hierarchy. In the National Forest System branch, these are the regional foresters, the forest supervisors, and the district rangers—the line officers, in

Forest Service parlance.

The Forest Service consists of three branches, reflecting the responsibilities assigned to it by the Congress. The National Forest System branch (with which this report is principally concerned) has responsibility for the management of the national forests. State and Private Forestry is, as its name indicates, responsible for federal forestry assistance programs to nonfederal agencies and private landowners. The Research branch conducts research into all aspects of forestry and makes the information available to state and federal agencies and the public.

The National Forest System, the largest branch, has 16,192 of the Service's 19,590 full-time, permanent employees.[28] Its administration is highly decentralized, with the nation divided into nine forest service regions. Two regions—8 and 9—cover nearly the entire area east of the 100th meridian. (See map, pages xxii-xxiii.) Region 9, the Eastern Region, with headquarters in Milwaukee, Wisconsin, is responsible for the 17 forests comprising about 11,300,000 acres in 22 states in the northeastern quadrant, spanning the area from Missouri to Maine. Region 8, the Southern Region, with headquarters in Atlanta, Georgia, has responsibility for 33 national forests of about 12,400,000 acres in 11 states, stretching from Texas to Virginia (the region also includes Puerto Rico and the Virgin

Islands). In a business sense, a forest region is no small enterprise. During the 1976 fiscal year, the Southern Region's budget amounted to $90,400,000 and that of the Eastern Region, $57,800,000—in the medium range as regional budgets go, but dwarfed by the $173,000,000 budget for the Pacific Northwest Region which covers only two states, Oregon and Washington.[29]

Organizationally, each regional headquarters is structured much like the Washington office. The regional forester is assisted by deputy regional foresters and a staff of specialists in the various forest activities, such as land acquisition, timber, wildlife, recreation, and various administrative functions. Regional staffs are large, with more than 300 employees in the Southern Region and about 220 in the Eastern Region. The regional forester is responsible for shaping Washington administrative policy and goals to the capacities of the individual forests in the region, for encouraging maximum performance from the forests' staffs, and for transmitting needs and problems from the forests back to Washington. Working with fund requests from individual forests, the regional forester assembles a budget proposal and transmits it to the Washington office. When the final budget allocation is issued from the Washington office, along with the chief's program directions, the regional forester distributes the funds to the forests along with his own direction statement establishing program priorities for the fiscal year. An important responsibility of the regional forester is monitoring the performance of the staffs of individual forests to ensure that their work follows the regional direction.

Again, the organization of a national forest headquarters is much like that of the regional headquarters. The forest supervisor, who may have a deputy, is assisted by staff specialists in timber management, recreation, and other program areas. Forest headquarters' staff size and budget vary widely, depending both on forest size and the intensity of various activities. (For instance, forests of high timber productivity have commensurately larger timber-management staffs.) In the Eastern Region, the smallest forest, the Green Mountain, has a permanent full-time staff of 56 persons, with 34 in the supervisor's office and the rest in the field. The Mark Twain has a permanent full-time staff of 156, with 63 in the supervisor's office. The 1976 fiscal-year budget for the Green Mountain was $1,480,000; for the Mark Twain, $4,700,000.[30]

The in-the-field operational unit of the National Forest System is the ranger district, within which the district ranger carries out policy and programs under the direction of the forest supervisor. Kaufman, in his examination of the work of the ranger, concluded: "In the last analysis, the elaborate overhead structure of the Forest Service has as its purpose controlling the behavior of these men who handle the real property of the agency, and who are in most frequent and often close contact with the public. Their tasks require the exercise of judgment; with 792 men [the number of district rangers at the time] making judgments critical to the execution of policy pronouncements, such control is not easily exerted."[31]

The National Forest System branch is primarily concerned with the management of federally owned land. Even though their activity often influences what happens outside forest boundaries, staff members of the National Forest System branch tend to step gingerly outside the national forests, for nonfederal land is considered the terrain of State and Private Forestry. This branch administers and manages numerous federal forestry programs that affect nonfederal land. These include financial assistance and technical aid to private landowners, insect and disease protection (a S&PF activity on national forests), and an array of activities categorized by the Forest Service as "rural development," which includes technical assistance to state and local agencies in such fields as leadership development, job training, and community services. S&PF also is responsible for preparing a Forest Service program to encourage consideration of forest values in land-use planning efforts of state, regional, and local units of government. It devotes considerable effort to assisting and strengthening state forestry agencies and regards the state foresters as its main constituency.

The division of responsibility between the National Forest System and State and Private Forestry has significant implications for the coordination of national forest management with land uses on inholdings and adjacent private lands. Although assistance to private landowners was the first activity initiated by Pinchot when he became head of the Division of Forestry in 1898, administration of the national forests came to dominate other Forest Service activities following their transfer to the Department of Agriculture in 1905. Increasingly, the National Forest System branch gained prestige, and State and Private Forestry came to be regarded as its poor brother.

In the West, where the national forests dominate the forest land base and opportunities to assist private forest landowners are lim-

From 1975-77, I was Commissioner of Environmental Management for the Commonwealth of Massachusetts, which is a state agency with many responsibilities— in land and water resource management, protection of wetlands and floodplains, and forestry. Our forestry programs were primarily concerned with the management of state land for high productivity, but also involved technical assistance to private landowners to maximize the state forest production.

The agency has strong ties with the Forest Service. In a number of state program areas, the federal cooperative forestry program was an important element; it is promoted by the Forest Service through the state foresters— and I was technically the state forester.

BETTE WOODY, Boston University
Former Massachusetts Commissioner of Environmental Management
Boston, Massachusetts

ited,[32] State and Private Forestry falls under the responsibility of the regional forester. But in the East, where S&PF was separated from the National Forest System in 1965 in order to give it greater visibility and independence, S&PF activities are the responsibility of area directors—the S&PF equivalent of regional foresters.

Recent Forest Service chiefs have tried to expand the responsibilities and upgrade the personnel of State and Private Forestry. In a 1972 speech, Chief John R. McGuire discussed the broadened role for S&PF:

... while State and Private Forestry activities were once rather limited in scope ... today, in addition to traditional cooperative programs with the state forester, our S&PF people are also charged with an increasingly broad range of federal responsibilities—and a wider diversity of coordinating roles with other federal and state agencies. The Congress, the President, the Secretary of Agriculture, and in turn, the Chief of the Forest Service have charged or delegated to the State and Private Forestry arm a large number of important responsibilities.[33]

S&PF personnel are proud of their distinctive role. "We now have an identity we didn't have before," commented one S&PF official about the branch's separation from the National Forest System in the East. Philip Thornton, deputy chief for State and Private Forestry, acknowledged past weaknesses of S&PF, but said, "Now we have the programs and the people—all we need is firm direction from Congress in the form of appropriations."[34]

The removal of S&PF from the jurisdiction of the regional foresters in the East was controversial at the time, with opinion divided within the Forest Service. Today, the scars of the controversy are still tender. Some doubts about its efficiency remain. An in-service study of Forest Service organization in 1976 cited problems of communication and coordination between the National Forest System branch and S&PF in the East. However, Chief McGuire and his top advisors reaffirmed the present arrangement, finding that it was working well and that communications problems could be resolved without reconsolidation.

While the initiatives being developed by S&PF in forest land-use planning and in assistance to private landowners are admirable, far greater cooperation is needed between S&PF and the National Forest System. Although examples of cooperation do exist, there seems to be little day-to-day contact in the field between S&PF and forest supervisors or district rangers. One S&PF official confided that "perhaps there is too much separation. Sometimes we forget that we're all part of the same organization."

Whenever possible, S&PF specialists in cooperative forestry programs and land-use planning should be based in forest supervisors' offices—though not under the direction of the forest supervisors. Similarly, S&PF programs of planning and technical assistance should make greater use of National Forest System staff. The forest supervisor should be given a direct role in reviewing—and promoting—S&PF assistance to owners of inholdings or private lands immediately adjoining the national forests.

PERSONNEL POLICIES

Those responsible for day-to-day management of the national forests represent a wide range of resource professions. Once the bastion of the generalist forester, the Forest Service now includes increasing numbers of civil engineers, hydrologists, soils scientists, wildlife biologists, geologists, landscape architects, and others whose specialized skills complement and reinforce those of the forester. Within the forestry profession, specialized areas have developed—in wildlife, recreation, and economics, for example.

It was once possible for federal foresters and their professional associates in resource management to concentrate on the management of the public land without paying much attention to what took place on surrounding private land. Usually, its use was compatible with the national forest. But times have changed.

In the East—and in some cases in the West as well—dramatic change is occurring on private land in and around the national

The Forest Service is a political animal, but many in the Forest Service really care. They are a high-grade bunch of people. The trouble is that they have gone to schools where there was a timber orientation. Consequently, the Forest Service has tended to overemphasize timber. They put together a timber-management plan and then hang the other things on the end of it. That's oversimplifying it, of course, but that's what it looks like.

Because the foresters work in the woods, they usually come to realize that there are other values beyond timber. But the schools that teach them have to change, and I think that's happening. Also, the political climate has to change so that people will stand up and say, "We want other values considered."

LUCY SMETHURST, Chairman of the Board, The Georgia Conservancy
Atlanta, Georgia

---

forests. Decisions made daily by state, regional, and local units of government affect forest management. In some cases, such as approval of a rural subdivision on a forest inholding, the impacts are immediate and obvious. In other cases, the decision may require highly technical analysis before the impacts are known. The siting of an interceptor sewer or the construction of a large waste-water treatment plant to serve communities near a national forest may open areas on the fringe to new residential development and bring new neighbors who may object to trail construction or timber cutting; a decision to build a new road or highway near the forest may have an impact on the amount of traffic on forest roads, the water quality of forest streams, the use of forest facilities, and the amount of litter and trash left behind.

In order to anticipate future demands upon the forests themselves, as well as pressures upon their boundaries, forest managers need to understand community development trends, the dynamics of land development, economic, social, and cultural changes in the region, and land-use regulatory techniques which can be employed by various levels of government. The enactment of new federal environmental laws which delegate to the states responsibility for air and water quality and require certain control over the uses of land has pushed the states into regulatory areas of immediate Forest Service interest. This requires that the Forest Service acquire new expertise in a range of professional areas beyond those dealing directly with resource management.

But so far, the response of the Forest Service to this new need has been to place resource professionals already on the staff into newly created positions which require specialized knowledge of planning techniques, state and local government operations, housing, employment, and transportation. The emphasis that resource professionals tend to place on the forests in isolation could be to blame for the Forest Service's inadequate attention to external influences on their lands, particularly the uses of adjacent private land. It also might explain the absence of a comprehensive evaluation of public forest-private land interrelationships.

When I was growing up, forestry meant dealing with the timber and the trees. Obviously now it's people, and forestry has to come along and mature and grow into that too. Sociological problems are much more important than the timber. We can grow the timber! I heard a senior forester, who now heads the University of Vermont School of Natural Resources, say the other day that if U.S. forestry's technical knowledge was suspended at the 1950 level, we could grow enough timber for the United States today with no problem at all. What we need now is people knowledge.
BRENDAN J. WHITTAKER, Chief, Information and Education
Agency of Environmental Conservation
Montpelier, Vermont

Specialists in fields other than the traditional resource professions are needed to integrate forest resources planning and management with local, regional, and national needs. **The Forest Service should recruit economists who understand regional economic interactions; urban and regional planners with expertise in nonforest land-use planning, employment, housing, and transportation; and lawyers and others skilled in mediation.** These additional professionals should be assigned to staff positions in regional offices and the Washington office to help prepare forest land-management plans and develop the Forest and Rangeland Renewable Resources Program. It would be their responsibility to bridge gaps between Forest Service planning and state, regional, and local planning efforts. Their efforts should be in addition to, and not a substitute for, S&PF programs to improve resource-planning expertise among local units of government and multicounty planning agencies. **Further, the Forest Service should develop its own training programs to broaden staff planning expertise in nonresource areas.** These should be designed in cooperation with universities through the development of short-term special courses or the enrollment of selected personnel for graduate degrees in appropriate fields. Nor should opportunities for forestry schools to broaden and liberalize their curricula be overlooked.

Recruitment of people with these skills should not be at the expense of Forest Service personnel with expertise in resource management. The Forest Service needs all the resource specialists it now has and more. There is an urgent need for an increase in personnel which, over the past decade, has been held to unconscionably low levels by order of the Office of Management and Budget and acceded to by the Congress. In fact, the permanent, full-time personnel ceiling for fiscal 1976 was 1,702 positions *less* than the figure for 1966,[35] despite new Forest Service responsibilities for the review of wilderness areas, preparation of environmental impact statements as required by NEPA, the preparation of forest plans, and the development of the Renewable Resources Assessment and Program.

The addition of new skills to the Forest Service might be aided by reconsideration of the current Forest Service policy of promotion almost exclusively from within. Within the National Forest System branch, the main line operations staff is service-grown, each staff member having painstakingly worked his way up the hierarchical ladder from an entry-level apprentice position to district ranger, through forest staff positions to forest supervisor, and so on. The Forest Service has estimated that during a recent 10-year period, fewer than 10 persons from outside the agency were appointed to a National Forest System line officer position or Washington office senior staff position.[36]

The Forest Service should give serious consideration to modifying its current promotion policy in order to allow entries to the Service at higher levels. This modified policy should be applied on a highly selective basis, primarily to staff positions rather than line ones, and should help attract specialists in nonresource professions. Changes in the current system would require a reworking of the Forest Service and Civil Service employment and promotional procedures and the definition of new job positions in forest headquarters, regional offices, and the Washington office. But such administrative inconveniences might well prove justified in view of the additional dynamism within the Forest Service that the infusion of new ideas and approaches would engender.

The Forest Service's policy of frequent personnel transfers also can affect forest management. In general, Forest Service professionals are transient workers, moving every few years, often as part of the training and promotional process, sometimes not. As Kaufman observed: "The Service does not merely wait until vacancies occur; it shifts men to replace each other in what looks like a vast game of musical chairs, but for the serious purpose of giving them a wide range of experience in preparation for advancement to positions that require a broader understanding of national forest administration than can possibly be gained in long assignments at a single duty station."[37] Also, transfers are a way of ensuring that the Forest Service professional does not develop strong local attachments that jeopardize his ability to make decisions with the national—rather than local—interest foremost. Often these transfers shift personnel from one forest to another within a region; sometimes they are more radical interregional moves, from western mountain forest to an eastern Appalachian forest, for example. "We try to pick the best available man for the job, whoever and wherever he may be," said Raymond M. Housley, an associate deputy chief for the National Forest System.[38]

Though the strictly administrative and resource skills necessary for managing national forests may well be readily transferable, problems are created by these frequent personnel changes. One former forest supervisor, who served for three and a half years in his old position before being transferred to a regional staff post, confided, "It took me nearly two years before I felt I was fully effective in my [supervisor's] job." Each forest operates within its own region's social, cultural, and economic environment. Contacts must be carefully cultivated among local officials, other community leaders, and user groups. The establishment of good working relationships with a community and a forest constituency takes time. Inevitably, local officials and user groups feel that frequent changes of forest supervisors break management continuity.

To eliminate some of this uncertainty, **the Forest Service should establish a policy of minimum terms for assignments—perhaps four years—in the main line positions** (district ranger, forest supervisor, and regional forester), and make this policy known to the public. Specific terms of service for given positions would inspire greater public confidence in both the official and the Forest Service, and foster public acceptance of staff change as a natural event in the course of national forest administration. However, the policy should be sufficiently flexible to allow the promotion of outstanding persons or the transfer of unqualified employees.

The changes proposed in Forest Service personnel practices are

not so extensive that they would undermine the essential structural stability of the service. Rather, they should help encourage the enlistment and use of people with fresh skills and attitudes who can help the Service develop new initiatives and seize opportunities now overlooked or too quickly discarded. Like other proposed changes, these are designed to give the Service the ability to deal with new and demanding problems.

New legislation, improvements in the Forest Service planning process, and changes within the organization to broaden its perspectives should provide the foundation for more sensitive management of the eastern national forests. Such changes should enable the Forest Service to work more cooperatively and effectively with its neighbors. This is essential if the national forests and the intermingled and adjacent private lands are to complement one another rather than coexist in an uneasy and even antagonistic atmosphere that is mutually damaging.

REFERENCES

Chapter IV

1. Organic Administration Act of 1897, Act of June 4, 1897, as amended (16 U.S.C. 473-478, 479-482, 551).
2. Weeks Law, Act of March 1, 1911, as amended (16 U.S.C. 480, 500, 513-517, 517a, 518, 519, 521, 552, 563).
3. Multiple-Use Sustained-Yield Act, Act of June 12, 1960, as amended (16 U.S.C. 528-531).
4. Richard E. McArdle in an address before the Fifth World Forestry Congress, Seattle, Washington, 1960.
5. Society of American Foresters, "Forest Policies of the Society of American Foresters," Washington, D.C., December 1973.
6. Daniel R. Barney, *The Last Stand* (New York: Grossman Publishers, 1974), p. 14.
7. Orris Herfindahl quoted in William A. Duerr, *TIMBER!: Problems, Prospect, Policy* (Ames, Iowa: Iowa State University Press, 1973), p. 45.
8. Marion Clawson, "The National Forests," *Science*, February 20, 1976.
9. U.S. Forest Service, *Forest Plan, Allegheny National Forest* (n.p., 1975), p. 47.
10. U.S. Forest Service, *Forest Plan, White Mountain National Forest* (Milwaukee: U.S. Forest Service, 1974), p. 51.
11. 16 U.S.C. § 1601 (1974).
12. *Forest Plan, Allegheny National Forest*, p. 47.
13. 16 U.S.C. 1 (1974).
14. U.S. Forest Service, "National Forest Planning," 1976.
15. U.S. Forest Service, *System for Managing the National Forests in the East* (Milwaukee: U.S. Forest Service, 1970).
16. U.S. Forest Service, *Guide for Managing the National Forests in the Appalachians*, 2nd Edition (Atlanta: U.S. Forest Service, 1973), p. 18.
17. U.S. Forest Service, *Forest Service Manual*, Title 8214—Forest Land Use Plan—1973.
18. Forest Service Manual, Title 8213—Major Planning Elements.
19. U.S. Forest Service, *Plan for Managing the Ocala National Forest* (Atlanta: U.S. Forest Service, 1972), p. 51.
20. Testimony of Senator Hubert H. Humphrey in Hearings on the National Forest Management Act before the U.S. Senate Subcommittee on Environment, Soil Conservation, and Forestry of the Committee on Agriculture and Forestry, November 20, 1973, 93rd Congress, 1st Session. (Washington: GPO, 1974), p. 2.
21. U.S. Forest Service, *A Recommended Renewable Resource Program* (Washington: U.S.F.S., 1975), p. 2.
22. Marion Clawson, "Assessment and Program of the Resources Planning Act," 1976.
23. Herbert Kaufman, *The Forest Ranger* (Baltimore: Johns Hopkins Press for Resources for the Future, 1960), p. 64.
24. Henry Clepper, *Professional Forestry in the United States* (Baltimore: Johns Hopkins Press for Resources for the Future, 1971), p. 66.

25. Robert E. Wolfe, "The National Forests: Background and Issues" (Washington, D.C.: Congressional Research Service, 1973).
26. From Glen O. Robinson, *The Forest Service* (Baltimore: Johns Hopkins Press for Resources for the Future, 1975), p. 22.
27. Michael Frome, *The Forest Service* (New York: Praeger Publishers, 1971), p. 26.
28. U.S. Forest Service figures as of August 1976.
29. Source: Program Development and Budget Staff, U.S. Forest Service.
30. Ibid.
31. Kaufman, *The Forest Ranger*, p. 48.
32. The 1977 budget allocates more than twice as much money ($15,100,000 vs. $6,900,000) to major S&PF activities in the East as in the West, an indication of the relative importance of state and private assistance opportunities in the two regions.
33. John McGuire in an address before the National Association of State Foresters, September 1972, Portland, Oregon.
34. Interview with Philip Thornton, Deputy Chief, State and Private Forestry, U.S. Forest Service, February 1976.
35. Source: Personnel Management Staff, U.S. Forest Service.
36. Robinson, *The Forest Service*, footnote 45, p. 51.
37. Kaufman, *The Forest Ranger*, p. 176.
38. Interview with Raymond M. Housley, Associate Deputy Chief, U.S. Forest Service, August 1975.

CHAPTER V

opportunities for cooperation and coordination

Friction between the stewards of the national forests and owners of adjacent private lands is inevitable, for it is always difficult to reconcile national needs, landowner attitudes, and marketplace realities. But there is much the Forest Service can do to increase the possibility that the uses of private lands will not undermine management plans for public lands and instead, ideally, will complement them in meeting local, regional, and national needs. In a position paper on land-use planning, the Society of American Foresters has stressed that plans for public lands "should be coordinated with regional, state, and local plans. . . . The objectives of [public] land use planning can be frustrated if the activities on surrounding lands are not adequately considered."[1]

Blighting developments, often a visual cacophony of strip commercialism or tacky slum subdivisions, exist in almost every eastern national forest. The effects can build subtly and accumulate until they change the nature and character of large forest units. This phenomenon is particularly dramatic in the North Carolina Highlands area of the Nantahala National Forest, Massanutten Mountain in the George Washington National Forest in Virginia,[2] and sections of Wisconsin's Nicolet National Forest closest to Milwaukee. In the Ocala National Forest in Florida, "land developers, operating without control of zoning, subdivided tracts of private land within the forest and have sold thousands of lots. Most developments have been built to the lowest of standards without adequate access, utility, or sanitary facilities. Their uncontrolled growth has posed serious problems for the Forest Service and local government. . ."[3]

On the Ocala, privately owned lands inside the forest boundary are a major problem. There are 130 to 150 platted subdivisions in the forest. These vary in size from a couple of acres to 600. Some lots are occupied by permanent residents and people live on them all year-round. Others are used for weekend homes or hunting or fishing camps. People may not know just where their property boundary—or the adjoining national forest boundary—is. Often the private landowner uses the national forest as his backyard. He thinks he can use it for a dog pen, a storage area, a garden, or a playground for his kids—and that just isn't so. Some are misled by overzealous real estate salesmen, who imply to the purchaser that the Forest Service doesn't mind if they use the land.

Subdivided inholdings also create problems for local government. They have to provide these scattered parcels with a whole range of services: road access and maintenance, school busing, law enforcement, fire protection, litter collection and disposal, and power and telephone utilities.

WALTER A. GUERRERO, District Ranger, Ocala National Forest
Ocala, Florida

But the effects of land uses flow not only from private land to public land. National forest land use can have a profound effect on nearby private landowners and communities. Development or activities in the national forest can increase or decrease the value of adjacent private land. The establishment of forest campgrounds can affect the volume of traffic on local roads and the need for improvement or maintenance. Recreational attractions of national forest land, such as ski slopes, whitewater rivers, and wilderness areas, can create opportunities for complementary uses on nearby private land. While areas designated for low-intensity use, such as wilderness and primitive areas, often have been considered to be of little economic benefit, increasing user interest in these areas may make them magnets for urban tourists. National forest clearcutting in the Monongahela was considered damaging to the area's tourist industry. Then too, Forest Service resource management can help or hurt a county's treasury, depending upon whether revenue-producing activities are increased or decreased. Legislation

The counties in my district are fairly sparsely populated. Forest County has a population of only 5,000—about 11 people per square mile—which is the lowest of any county in the state. According to the state police, the population expands to 50,000 on weekends in the summer. Blue collar workers have been able to expand their weekends to include Thursday and Friday and/or Monday and Tuesday, so that Wednesday gets to be the only day of the week that the Allegheny National Forest isn't crowded. I get appeals for more men to be assigned to the police post because of such a tremendous influx of campers coming to the forest.

ROBERT KUSSE, Member, Pennsylvania State Legislature
Warren, Pennsylvania

enacted by the 94th Congress should insure some stability and increased payments to counties, thus mitigating the effects of fluctuations in national forest timber harvests. (See pages 225-231.)

Nearly every forest plan acknowledges the interdependence of public and private lands, and expresses a policy of cooperation with state, regional, and local units of government. Although many forest supervisors sincerely believe they are engaging in cooperative action, their approach tends to be so guarded that effectiveness is dulled. With a few exceptions, the Forest Service reacts to problems, rather than initiating action to realize opportunities.

The reciprocal impacts of land decisions constitute a compelling argument for close cooperation between national forest managers and local officials. The chief obstacle to cooperative action is conflict—real or perceived—between the expectations and desires of forest users distant from the forest scene and local economic aspirations. As one local official put it: "The goals of a region are quite different from those of the national forest. The constituencies are not the same." To say that mutual benefits will have to be recognized and trade-offs negotiated is an obvious oversimplification.

A number of other factors also tend to inhibit cooperative land-use management in forest regions. Many states are reluctant to assert control over land uses of greater than local significance. Rural areas generally tend to oppose land-use planning and regulation, and local officials are particularly bitter about perceived inequalities in the system of sharing national forest revenues with local jurisdictions. For its part, the Forest Service offers no program of cooperative action at the individual national forest level, and staff members hesitate to assume a leadership role in planning for

The town of Bartlett is almost 70 percent national forest. Many people argue that we therefore don't need zoning or subdivision controls—because there's already so much open space. People don't like to be planned for and told what to do—and that's not all bad—but we're left with no control over the quality of development we get.

JUDITH SMITH, President, American Association of University Women
North Conway, New Hampshire

---

a forest region.

Carl H. Madden, chief economist of the Chamber of Commerce of the United States, has pointed out: "More than 10,000 governments today regulate how land is to be used, though much land is under no zoning restriction."[4] Some states, alarmed by deteriorating land and other resources, have begun to reclaim the land-use regulatory authority previously delegated to local jurisdictions. Federal air- and water-pollution control laws and programs are also influencing rural-development patterns. And, of course, the Forest Service is developing its own land-management plans for individual forests as well as the national program required by the Forest and Rangeland Renewable Resources Planning Act. In many instances, one agency is only dimly aware, if at all, of the activities of the others. The challenge, in forest regions and elsewhere, is to bring order to this chaos.

Within and around most national forest areas, unsophisticated rural communities—localities often without adequate technical or financial resources — are confronting difficult and increasingly complex land-use problems. In many of these counties, a national forest constitutes the largest block of land in single ownership.

In a technical sense, the New England River Basins Commission has no authority to impose any decision on a state, local government, or individual federal agency. Several different agencies at each level of government —federal, state, local—forage back and forth over a resource base, each with its own view of how that resource ought to be managed, and each with authority for some kind of management action. If you can get the various agencies to plan what they're going to do in concert with each other, perhaps you can avoid conflicts, be more sensitive to both environmental and economic concerns—and save money, too. And that's the theory behind the Commission.

But the institutional mix presents an unbelievably complex situation. The system by which governments manage resources has grievous flaws in it. You've got three different levels of government, and every interest group operates at each level: all the development interests and all the preservation interests are active at the federal level, the state level, and the local level. And the political jurisdictions served by the various governments don't coincide with the boundaries of the natural systems—in our case, the river basins. It's a nightmare, and it is the central dilemma of natural resource management in the United States.

The Commission is able to see all sides of natural resource issues because all the parties at interest are represented. It's the nature of our society that there is little political support for this holistic view. Everybody lines up and argues a point of view vigorously with the minimum latitude for compromise, and the balance is struck through the political process rather than through the analytical process. The Commission tries to strike a balance through the analytical process.

R. FRANK GREGG, Chairman, New England River Basins Commission
Boston, Massachusetts

This fact, combined with Forest Service expertise in resource management and its authority to provide technical assistance to local jurisdictions and landowners, offers an extraordinary opportunity for the forests to catalyze land-use planning and coordinate various existing programs.

But cooperation is hampered because local officials and citizens feel the Forest Service is unresponsive to their needs. For the most part, the Forest Service has remained apart and resisted intrusion into its management domain. "Sure, they call us all together and ask us what we think," said one local official, "but when the decision is being made, they slam the door and hang up a 'don't disturb' sign." In return, local units of government are rarely sensitive to the impact of private land uses on national forest resources. "Why should we worry about the national forest—it doesn't pay taxes," is all too common a view among county officials.

Additionally, rural residents tend to distrust land-use controls. One common explanation was expressed by an observer: "Local land-use control doesn't work because long-range benefits do not provide enough incentive to get a politician reelected." Rural governments, often fiercely independent, rarely see a need to plan regionally or to be concerned with issues, such as water quality, that range across jurisdictional boundaries. And even in areas where land-use laws have been enacted, their effectiveness is questioned.

Since so much of our land is in the national forest, we don't have much that's developable. We talk to the Forest Service far more than the state planners.

I think there was a serious lack of planning when the forest was established and land acquired by the Federal Government. There is a single road through the national forest between Ripton and Goshen. There are only six private parcels on the road; no one lives there year-round, yet the road has to be maintained. There is a camp up there with a caretaker with two children. We have to send a school bus up there, and we don't get a dime back on it.

But on the other hand, I may be painting too dark a picture. Relations between Ripton and the Forest Service are pretty good. The Forest Service has helped us repair bridges when the bridges provided access to a timber sale, and we've helped them acquire rights of way for access to national forest land. They need us because they have large holdings in the town; and because they're in the town of Ripton, we have to work closely with them.

RICHARD CLARK, Selectman
Ripton, Vermont

LOCAL, REGIONAL, AND STATE PLANNING PROGRAMS

Efforts to enact even basic zoning and subdivision development regulations, common in most urban areas, meet with failure as often as success. Of the nine West Virginia counties that encompass the Monongahela National Forest, none had subdivision controls, and only two had even partial zoning in 1976.[5] Most forests operate in a slightly better regulatory milieu: In Virginia, 16 of the 34 counties in which the George Washington and Thomas Jefferson National Forests are located had zoning ordinances by the end of 1975.[6]

But some localities do give special consideration to their forest resources. Warren County, Pennsylvania, for example, enacted a zoning ordinance which put much of the private land in the Allegheny National Forest in a conservation zone, although its one-house-per-acre restriction still permits considerable development.[7] Some counties, such as St. Louis County, Minnesota, have a zoning category for forest land. Even so, special zoning for forest environs remains the exception.

Since decisions involving land, water, and air commonly affect more than one county, one might look to regional authorities to coordinate land-use and resource planning. But multi-county regional planning agencies, while common in the East, are still in their infancy. In most instances, these councils' activities are confined to those regulated by participating governments, often activities required by federal or state legislation. Few have the wholehearted support of the participating jurisdictions. The fact that participation is still voluntary inhibits the making of tough decisions which might ruffle a key jurisdiction. In Florida, a supervisor from the largest county in a rural regional council said, "Polar bears will walk the streets of Miami" before the council would disapprove a project his county wanted.

But some regional progress also is being made. In Virginia's Mount Rogers Planning District, a regional planning area that encompasses five counties around the Mount Rogers National Recreation Area of the Jefferson National Forest, district planners have provided valuable assistance to local jurisdictions, only one of which has a planning staff. The district's recreation plan attempts to mesh private land uses with the "rural America" theme of the Mount Rogers National Recreation Area.[8] In northern Minnesota, site of the Superior National Forest and its Boundary Waters Canoe Area and also the Voyageurs National Park, the Arrowhead Regional Development Commission is helping counties direct development attracted by the new park.[9]

In many areas of the country, multi-state river basin commissions are preparing water-resource plans for large regions, with the Forest Service's State and Private Forestry branch providing the forest element for no fewer than 67 river basin studies.[10] However, river basin planning is progressing slowly and the quality of work is uneven. Funding has been a problem, too. As with most regional planning efforts, there has been a reluctance to confront the difficult issues. Further, since the national forests comprise but a small portion of the larger area, forest issues and opportunities tend to get shuffled under issues that more directly affect urban regions.

Some states are responding to the problem posed by ineffective land-use controls at the local and regional levels.[11] In 1970, then-Governor of Vermont Deane C. Davis, alarmed over plans for a development that would have created a "new town" with a population equal to that of the state's third largest city, appointed a commission to consider the problem of recreational home development.[12] The study resulted in Act 250, a statewide land-use control law, which has benefited the Green Mountain National Forest. Florida enacted a comprehensive land-use law in 1972,[13] and this

The River Basins Commission's responsibility runs to "water and related land resources." Now that phrase encompasses virtually all natural resources. In that context, the forests become a major physical feature of the region and are appropriate subjects for the kind of multiple-purpose planning and management that we're trying to apply to the rivers. You're talking about the forest as a high-quality recreation resource, as a producer of fiber, of wildlife habitat, of wild land generally, and as an employment opportunity. We have explored the possibility of funding a study—a detailed analysis of the relationship between water management and forest land management. We would look at the degree to which water management considerations affect the efficient economic utilization of the northern New England forest, and vice versa. Do we run some risks that controls imposed on forest utilization in the interest of water resources—say from increased costs of preventing erosion in road construction, or restrictions on forest pest control—may affect the economic prospects for using forest resources? Maybe we ought to see whether we couldn't use the Commission as an instrument for thinking through water management issues and water-related land issues and look at those in the context of forest management. Perhaps we should try to come up with some policy and even specific program recommendations that would help make certain that both the water management and forest land management objectives made some sense.

R. FRANK GREGG, Chairman, New England River Basins Commission
Boston, Massachusetts

law may positively affect the national forests in that state. Among other provisions, it permits the state to designate "areas of critical state concern," which are to be protected by local or state action. Other states are showing an interest, but moving warily. The Virginia General Assembly enacted legislation in 1975 requiring that all local governments adopt subdivision ordinances by July 1, 1977, and prepare comprehensive plans by July 1, 1980.[14]

In addition to these local, regional, and state planning and regulatory approaches, other new techniques being tested could have significant benefits for the eastern national forests. Special programs designed to control development around particular areas or types of "critical" land—usually scenic, environmentally fragile, or historic land, or land with economic value if left undeveloped—are attractive because they can be identified with land possessing popularly recognized environmental and economic values. Perhaps the best known of this genre are land-regulatory programs aimed at controlling development in coastal areas. Virtually every coastal state, with the stimulus of federal financial aid, is preparing a coastal-zone management plan. Some states have enacted more comprehensive legislation encompassing a variety of different types of "critical areas," such as mountains, wetlands, historic sites, and even prime agricultural land. These state laws compel implementing action by local jurisdictions to regulate development. Such state initiatives could result in controls over the development of land in and around the eastern national forests. However, at the end of 1975, only 5 of 24 eastern states with national forests had enacted critical-areas legislation.[15]

Additionally, a number of states have enacted, or are considering, special programs to protect prime agricultural land, with special tax considerations for agricultural land as the prevailing avenue of attack. While productive cropland is the principle concern, the laws in 18 states extend eligibility to forest or timber land as well.[16] Some states have changed their approach to taxing timber land as real property in order to encourage timber growing by private landowners.[17] Such approaches include the imposition of sev-

I think forest landowners might find good reasons to support stronger land-use controls. Doesn't the present pattern of development of profit-making, tourist-serving recreation facilities in the forests tend to build into the forest the seeds of political discontent? The increasing number of people in the forest community interested in management of the forest for aesthetic objectives are pushing for tighter forest management practices; this could impair economic management of the forest resource. You wind up with a lot of forest and shoreland property owners—condominium owners—who will lobby against this or that. I use this argument to suggest that the forest products industry might have a real stake in a state land-use allocation system. They might help develop state land-use policies and practices that would protect the economic value of the forest.

R. FRANK GREGG, Chairman, New England River Basins Commission
Boston, Massachusetts

erance taxes which levy the major tax claim upon the timber at the time it is cut, rather than while it is growing.

Other experiments designed to protect agricultural land show promise for forest management as well. New York State's "agricultural districts" legislation, which amounts to a comprehensive strategy for farmland protection, is attracting growing interest; a similar proposal has been advanced in Virginia. In Connecticut and New Jersey, purchase of farmland or farm-development rights is under serious discussion. At the local level, Suffolk County, New York, has approved legislation authorizing the expenditure of $21,000,000 to purchase the development rights to farmland.[18] Such a program might also be applied to forest land.

In view of this maze of cooperative planning and regulatory mechanisms already in place or under consideration, how can the national forests best cooperate with other units of government within a regional framework?

Of necessity, national forest supervisors, district rangers, and other staff members maintain contact with local officials and state and regional resource and planning agencies. The type and formality of contact vary widely among forests and even among districts on a single forest. Contact can be as informal as a discussion over coffee between a district ranger and county commissioner or as formal as the signing of a cooperative agreement for the exchange of information and coordination of activities. In some areas, Forest Service personnel have been assigned to work in the offices of regional planning districts.

Yet despite frequent contact, the usual stance is arm's length. Local officials often are highly suspicious of Forest Service interest in their activities. And, in spite of an elaborate structure for eliciting local comment on national forest plans, Forest Service personnel stoutly defend their land-management prerogatives.

I think local people should determine how land is to be used. But in the case of the town of Bristol, Vermont, the Forest Service wanted to tell us what we could and couldn't do with our land. The issue was over the Bristol Cliffs wilderness, where about 60 people owned land. In April, 1975, they received a letter from the Forest Service notifying them that the President, in January, had signed a bill creating the Bristol Cliffs Wilderness Area. In addition to about 5,000 acres of land within the Green Mountain National Forest, the wilderness area included about 2,900 acres of private land.

Well, the Forest Service had a meeting and the people were almost in shock. I don't own land in the wilderness area, but I went to that meeting. It was the first time the people had heard of the wilderness plan. We found out that the Forest Service had held a public hearing on the proposed wilderness the year before in Concord, New Hampshire! I'll tell you, people cried that night. The Forest Service, a branch of our government, stood there with their maps of the wilderness area and told us "Here's what you can do with your land, and here's what you can't do. You can't cut a tree unless it's dead. You can't put a road in unless we approve it. You can't change your house unless we approve it." There were about nine things they couldn't do with their own property. But they could continue to pay taxes on it.

Well, I came out of that meeting and was approached by a young man, Joe Conowal, who owns land in that area. Tears were running down his cheeks and he said, "John, will you help us?" I said, "Damn right I will." So we organized the Bristol Cliffs Wilderness Area Landowners Association. We wrote to Congressmen, and kept the pressure on. We did a lot of work. And we won. Bristol Cliffs is one of the few wilderness areas that has been redefined. Congress passed a law redrawing the boundaries, taking out all the private land.

The Forest Service made a bad mistake in human relations. The wilderness area would have affected a lot of families severely. For example, there is a family which owns about 200 acres that were included in the wilderness. The family has cut timber each year to keep the elderly mother in a nursing home. Suddenly they were told they couldn't cut the timber any more.

Now, there is not a single one of us opposed to wilderness areas. If we were, we wouldn't be living here. And we all see the many advantages to having the national forest. It attracts people and business to the community, and Lord knows we need that. A camp outside of town, recreational use of the forest—nobody objects to that. The town will have to use controls so that we don't overbuild in terms of homes and trailer parks. We must use restraint. We don't mind people coming up here and being part of the community, part of the forest. But we don't want the government saying we shall, or will, or must!

JOHN A. HISE, Member, Vermont Legislature
Bristol, Vermont

SOME EXAMPLES OF COOPERATION

Nationally, the most notable examples of cooperative relationships between the Forest Service and state and local units of government are in the West, where the forests have long been in public ownership. There, for instance, agreements have enabled personnel in the Teton National Forest in Wyoming and the Beaverhead National Forest in Montana to help counties prepare comprehensive plans and review and comment on proposals affecting these forests.[19] But such cooperative agreements are the exception in the East, where aggressive acquisition programs can abrade Forest Service relationships with local officials and citizens. A Conservation Foundation survey of eastern national forests in early 1976 revealed formal agreements between forests and local planning agencies only on the George Washington National Forest in Virginia and on forests in North Carolina.

More frequent are ad hoc arrangements to solve a specific, common problem. This cooperative approach works best when the Forest Service is in a position to make land available for a facility desired by local officials. Thus, the Nantahala National Forest provided a site for a county landfill, and incidentally helped solve a litter problem on the national forest.[20] Understandably, cooperation is more attractive to local governments if local benefits—especially monetary savings—are easily recognized. But the benefits of joint planning usually are not so obvious. It is a substantial jump from closing dumps to the development of a comprehensive plan for a forest region. In most instances, the Forest Service moves circumspectly in its dealings with local governments. "We try to tell them what we think," said one Forest Service official, "but we do it subtly."

One of the best examples of Forest Service cooperation with state and local governments is on Vermont's Green Mountain National Forest. There the staff has moved beyond the rhetoric of its policy to "cooperate with State and Private Forestry to encourage land-use planning at the state and local level."[21] Under an agreement with the State of Vermont, the Green Mountain National Forest staff participates in the state's land-use control process under Act 250.[22] With the county forester, Green Mountain district rangers inspect sites near forest land proposed for development and offer their comments. Frequently, the forest supervisor's office provides technical information to the District Environmental Commission, which decides whether the development is to be permitted. At the planning level, the forest staff is working with local units of government to plan for land uses on private land inholdings, future land adjustments, rights of way, access roads on national forest and private lands, present and potential zoning, local tax structures, and other aspects of the planning process.

The staff's understanding of national forest impacts on adjacent communities is reflected in its land-use plan for the Deerfield River Area, the site of highly developed ski resorts which use forest land. Under the Forest Service plan, the expansion or development of three area ski resorts which use national forest land will be approved only if the development will "not place a significant impact on available community services and facilities."[23] The national forest further pledges to "maintain a close liaison with the State Planning Office, the Agency of Environmental Conservation, and Regional and Town Planning Commissions so any proposals for development within or adjacent to the national forest may be assessed in relation to overall effects to the forest and its management direction, as well as to the local communities."[24]

The Green Mountain National Forest has actively sought public involvement in its land-use planning process as a matter of necessity, for Vermont's legislation permitting the Federal Government to purchase land for national forests requires that acquisitions be approved by town selectmen. Over the years, a candid relationship has developed between the public, local officials, and the Forest Service, encouraging cooperation rather than antagonism.

We're a comprehensive regional planning agency. Most of our time is spent directly with member towns in technical assistance—town plans and town land-use controls. We're also the review agency (OMB-A-95) for any federal action. That's how we have come in contact with the Forest Service recently. Most of our contact with them is reactive. They send us a plan for review, and we react to it. However, a town may ask that we get involved in some business involving the Forest Service.

Aside from periodic reviews of national forest plans and requests for information, we don't have an on-going involvement with the Forest Service. Forest Service officials are very good about responding to specific requests for information. But sometimes we feel a sort of gap or lag because we don't know what kind of planning process they're conducting. For example, we're doing several land-resource inventories for towns that have substantial amounts of Forest Service land. Usually we aren't concerned about the capability for development of national forest land because it's in Forest Service ownership. However, we've learned recently that the Forest Service has been studying private land in some detail when those lands are within the forest proclamation boundary. We'd very much like to be brought into that process, or at least to benefit from that information, when it's gathered. I think the Forest Service should be more actively involved in local and regional land-use and environmental planning.

In the case of the Deerfield River area and its ski facilities, I was rather impressed by the Forest Service's impact study. I thought that they had looked at most of the sources we would have used—local town plans, regional plans, state plans. And they made a real effort to address the environmental and economic issues. The impact of Forest Service plans on a locality should be of prime consideration. For example, I think it would be wrong for the Forest Service to decide—without any local input—that a particular area was going to be designated for development.

WESLEY WARD, Planner, Regional Planning and Development Commission Wyndham County, Vermont

Assigning Forest Service personnel to a local government or regional planning agency has also proved an effective means to increase cooperation. An example to be emulated—although Forest Service officials cite it as an "unusual situation"—was the one-year assignment of a Forest Service forester to the Arrowhead (Minnesota) Regional Planning and Development District (ARPD). The forester had previously served in a number of positions on the Superior National Forest, which lies within the Arrowhead District. With the ARPD, he worked on the preparation of a comprehensive plan for the Voyageur Planning Area, a 2,000,000-acre area on the periphery of the new Voyageurs National Park and adjacent to the Superior National Forest.[25] According to ARPD officials, his work "helped ensure that the planning effort recognized the importance of forest resources and forest management."[26] Local officials were persuaded that the timber resources on the periphery of the national forest were of such economic value that they warranted protection from development. One of the counties involved later enacted a zoning ordinance that included a "forest division"—an area designed for forest conservation. The success of such cooperative efforts, of course, still depends finally on the effectiveness of local and regional planning and local land-use regulation.

TOWARD IMPROVED COOPERATION

At present, the Forest Service has no coherent organizational framework for its cooperative efforts. Activity is divided between two branches of the Forest Service—the National Forest System branch, which administers the national forests and deals directly with local jurisdictions and landowners in the national forest planning process; and the State and Private Forestry branch, which has responsibility for providing forestry assistance to state, regional, and local agencies and landowners.

Currently, attention to cooperative programs and agreements in Forest Service policy documents focuses on the State and Private Forestry branch; there is little reference to cooperative opportunities for the National Forest System, nor direction to forest supervisors to engage in cooperative planning. Both the National Forest System branch and State and Private Forestry have land-use planning roles in the private sector and should work closely together. National Forest System staffs have a responsibility to participate in local planning efforts as one of the principal land managers in the area. State and Private Forestry should assist with cooperative funding and coordination at the state forester level and provide technical assistance at the local planning level.

The Conservation Foundation recommends that the State and Private Forestry branch develop, with the cooperation of the National Forest System, a Forest Service cooperative action program specifically directed at areas in and around the national forests. This program should organize the Forest Service's existing, diffused programs relating to rural development and cooperative planning into a service-wide, intergovernmental cooperative program. This action program would consolidate and coordinate existing Forest Service programs now under the jurisdiction of the State and Private Forestry and National Forest System branches, as well as related programs administered by other federal agencies. These should include technical assistance to local governments and landowners, public facility development, cooperative agreements, special-use permits, water-pollution control programs, water management for flood control and recreation, and manpower programs.

- The cooperative action program should be spelled out in the Forest Service Manual, with specific direction to forest supervisors on cooperation with local officials, including specific programs and agreements, and published in the *Federal Register.*

- A forest-level cooperative action program, describing steps the Forest Service plans to take, should be a required element of each forest plan.

- Forest Service governmental assistance programs should be targeted to areas near national forests where there are especially sensitive land-use problems.

- Where local officials are receptive and where forest resource values and land ownership patterns warrant it, demonstration cooperative programs should be initiated to test innovative approaches.

- The Forest Service should be alert to, and call attention to opportunities for, use of national forest land and adjacent private lands in ways that are complementary.

If cooperative land-use planning and management activities are to be instituted and effective, the Forest Service must abandon its traditional responsive posture. Where no other agency is prepared to take the lead, the Forest Service should do so. This does not mean that the Forest Service should dominate the regional plan-

The area's changed a lot. Different kinds of people are coming in from everywhere. We've made a lot of new friends. We remember when there were no cars, no highways here. We lived just four miles from Boone. There weren't a dozen houses in Boone then. We walked there. Didn't have to buy much except sugar and coffee.

MR. & MRS. CLARK ISAAC
Violis, North Carolina

---

ning and regulatory process, but it should assume the initiative in defining problems and opportunities from the perspective of the national interest.

The Forest Service role suggested above will be feared by some local officials and citizens and resented by others. The Forest Service will also be apprehensive. One Forest Service official commented that "this is a sure way to stir up big trouble with the local folks." But it seems evident that in many rural areas, local leaders will respond to the opportunity for cooperative action if mutual benefits are identified constructively. A Georgian offered this counsel in approaching local governments: "Show them how they can profit and they'll believe you; tell them it's their duty and they'll be contemptuous."

In cases where a local unit of government, a regional planning and development agency, or a state or multi-state agency has the interest, will, and capability, it is entirely appropriate that one or more such agencies assume or be assigned the lead role for forest-region planning. There may be situations in which a state wishes to designate a forest region as an area of state concern and establish a structure for integrating the management systems of affected governmental units. Or a group of local governments already united in a regional planning and development district might wish to give more coherence to their programs and take advantage of Forest Service assistance. A river basin commission or regional development commission might want to develop a comprehensive plan for a forest area, integrating the programs of all the affected governments. Forest Service personnel should encourage and cooperate with such efforts.

Mutually supportive and beneficial results also can flow from Forest Service cooperation with states over the increasingly critical problem of extraction of privately owned minerals beneath the forest surface. Their removal often is beyond the direct control of the Forest Service. (See pages 57-60.) In cases where the ownership of the minerals is reserved, the only authority for regulation of mineral development rests with state agencies.

Although federal strip-mine legislation failed to pass the Con-

Drilling for oil near the Allegheny National Forest in Pennsylvania.

Surface mining for coal on the Ironton District in Ohio's Wayne National Forest in 1954. While this land was restored and is now forested, the surface wounds from some old mining operations on the Wayne have been slow to heal. Present Ohio law now requires a high standard of reclamation.

Fumes from copper smelters have eradicated the vegetation in this Tennessee valley between the Nantahala National Forest in North Carolina, the Chattahoochee in Georgia, and Cherokee National Forest in Tennessee. The forested mountains of the Cherokee are in the background.

> The only way we can get anything accomplished to control oil drilling on the Allegheny is through cooperation with the surface owner. Our track record has been a lot better in recent years. If we can't reach agreement, we can turn to the state for assistance—they're usually close at hand anyway, watching for pollution and soil erosion—and there seem to be a lot of teeth in state laws.
> ROBERT L. FIELDS, Lands Staff Officer, Allegheny National Forest Warren, Pennsylvania

gress, some states have enacted laws to control this most destructive of mining techniques and to require the rehabilitation of disturbed land. Federal and state water-pollution control laws, if enforced, can require mining operators to protect water resources.

In Ohio, which now has a stringent law for controlling surface mining of coal, the Wayne National Forest largely relies on the state to ensure that those removing privately owned minerals keep surface disruption to a minimum, control water pollution, and make sure that the land is ultimately restored.[27] On the Allegheny, where there is intensive oil drilling, state environmental laws offer the only protection of forest resources on 260,000 acres on which rights are reserved.[28]

The Wayne and Allegheny, along with the Daniel Boone National Forest in Kentucky, provide excellent examples of national forest and state agency cooperation to control mining. Under a cooperative agreement with the state, the Wayne National Forest and the State Department of Natural Resources jointly administer licenses to surface mine for coal on national forest land. Applications for state licenses are sent to the forest supervisor for review and comment, and Forest Service recommendations are written into the licenses. While this arrangement has not solved all the problems of mining on the Wayne, it has helped significantly in the control of the most serious environmental abuses and reduced the impacts of mining on other national forest resources.[29] A similar arrangement on the Daniel Boone has effectively stopped strip mining on the national forest.[30] On the Allegheny, national forest staff members work closely with the Pennsylvania Department of Environmental Resources to monitor oil operations.[31] There, the cooperative efforts of the Forest Service and the Department of Environmental Resources resulted in a $29,000 levy against the Pennzoil Corporation for oil pollution of a mountain stream.[32]

We've been working on the national forest since 1956. We've never experienced any serious problems with the Forest Service, or any surface landowners for that matter. When we go into areas we mulch, seed, and terrace the cleared land and put in sluicing and ditching to control erosion and run-off and protect the surface. In the case of the Forest Service, we're dealing with people who are professional in their operation, who have a job to do to protect the forest and see that it's wisely utilized. They can help in laying out the oil operation—and they are also much less expensive than consultants.

We have to cut right-of-way and access roads. The Forest Service has made recommendations on road-building and siting, and we're following some of them. They've recommended wider main roads, beyond the size we need, so they can serve multiple purposes—as rights-of-way for power lines, for example.

The Pennsylvania Department of Environmental Resources brought to our attention that our oil drilling operations were in violation of the Erosion and Sedimentation Control Act. It was on the forest surface. We were asked to agree to avoid this in the future, police the area of the infraction, and make a donation to the state of Pennsylvania.

WARREN PAYNTER, District Engineer, Pennzoil Company
Bradford, Pennsylvania

FUNDING FOR COOPERATIVE PLANNING

Development of cooperative land-use planning programs has been a matter of very low Forest Service—and congressional—budget priority. In fact, there has been no specific budget item for such programs in recent years.

State and Private Forestry funds some comprehensive planning programs in rural areas on a "special project" basis through its assistance to state foresters.[33] In most of these cases, a state forester asks the Forest Service to pay a part of the cost of assigning one of his foresters to a state or regional planning agency to ensure that forest resources are considered in the planning effort. Since S&PF cooperative planning assistance projects are initiated by state foresters and not by the Forest Service, it is a fortunate coincidence if the area targeted for assistance by the state forester includes a national forest. Because of its reactive posture, S&PF does not deliberately target assistance money to areas near national forests.

The Forest Service's cooperative comprehensive planning statistics include the time spent on these activities by personnel of the National Forest System. Yet forest supervisors have no separate budget to support planning assistance or cooperative planning programs. Work conducted with local agencies must be funded from other budgeted activities, such as timber, recreation, and wildlife, and justified on the basis of benefits to those activities. This inevitably constrains the forest supervisor in his allocation of limited funds from direct resource management to more ephemeral planning and liaison activity. "I would like to enlarge our contacts substantially," said one forest supervisor, "but the money and manpower aren't there."

Because cooperative action and technical assistance are not budgeted activity items for national forests, much of the cooperative planning work by national forest staffs tends to be informal, as, for instance, in cases where national forest staff members work with local governments in the development of a land-use plan or zoning ordinance. Since relatively little direction is provided national forest personnel, the types of comprehensive planning projects undertaken are determined by the personal interest and initiative of a forest supervisor or district ranger, and not by major policy or management priorities.

The cooperative comprehensive planning programs of State and Private Forestry and the cooperative planning efforts of the National Forest System staffs both suffer because no funds are specifically allocated for those purposes. **The Congress should grant adequate and specific funding for State and Private Forestry to provide technical forestry assistance to state, regional, and local planning agencies whose decisions directly affect national forests. Within the National Forest System, forest supervisors should be allocated funds adequate to enable forest staffs to work with local officials and planning agencies on a continuing basis, providing staff assistance to the planning agency where appropriate.**

SPECIAL INSTITUTIONAL ARRANGEMENTS

The cooperative course toward the goal of complementary uses of public and private lands can be slow. It is risky, too, relying as it must on good intentions and informal agreement as well as official effort. Even stronger measures are necessary to achieve land-management coordination in cases where an area possesses resources of particular national or regional value which might be irretrievably lost if forthright action were not taken.

For the Cape Cod National Seashore and the Sawtooth National Recreation Area, the Congress determined that special controls were indeed required if these public areas were to be protected from unrestrained development of adjacent private lands. The New York State Legislature reached a similar decision in the case of the Adirondack Park. While each set of controls was developed as a site-specific response to a particular geographic and land-ownership situation and designed to accommodate the capabilities of local governmental units, some of the techniques employed may prove transferable, if used sensitively, to national forest land-management problems.

The Adirondack Park: Within the "Blue Line" drawn by the New York State Legislature when it created the Adirondack Park in 1892 lie 6,000,000 acres of land—one-fifth of the total land area of New York State. Of this 6,000,000 acres, 2,300,000 acres—38 percent—are state forest lands; the remaining 3,700,000 acres are privately owned.[34] As in most eastern national forests, private lands and established communities are interspersed through the state's forest preserve land.

The park has long been a summering place for New York City residents. In the late 1960's, with completion of the Northway—the New York-Montreal link which cuts the eastern edge of the park—automobile travel time from the city to the heart of the park was reduced from eight to four hours. And with the Northway came intensified pressures for development. A recent report by the Adirondack Park Agency recalls the situation as it existed in the late 1960's: "The result of the Northway and other highway improvements, of increasing leisure time and growing affluence, of an expanding population and a growing human need to get away from the congestion and pollution of metropolis, if only for a day or a weekend, posed a new problem for the Adirondacks. This problem was the threat of rapid, uncontrolled development that could, in the space of a generation or two, destroy forever the character of the region."[35]

Stimulated by a private move to have a portion of the Adirondacks converted into a national park, Governor Nelson Rockefeller in 1968 appointed a Temporary Study Commission on the Future of the Adirondacks. The commission, submitting its report in 1970, declared: "Park planning and land-use controls can best be implemented via an independent, bipartisan Adirondack Park Agency with parkwide control. Only through such a centralized land-use framework can state, regional, and local concerns all be provided for. The commission is confident that a productive partnership can be effected between the Agency, the Department of Environmental Conservation, local governments, and regional planning bodies. As local governments become more involved in the planning process, they should assume increased responsibility." [36]

The state legislature's decision to place local land use under the control of the Adirondack Park Agency has generated controversy. Yet it stands as the best example of an attempt to ensure protection of a priceless resource while providing local units of government the widest possible latitude under the circumstances. As the Adirondack Park Agency reported in 1976, "Two years' experience in administering the Adirondack Park Agency Act have shown that this unique law is basically sound, though within the Adirondack Park it remains controversial and widely misunderstood." [37]

Briefly, the Adirondack Park Act and its implementing Land Use and Development Plan provide for overall development densities for six different categories of land uses.[38] Decisions of nonregional impact are left to local units of government if the proposed project meets some basic land-density guidelines. The Adirondack Park Agency holds review and permitting responsibility for developments of regional impact, which are defined as projects of a certain magnitude or projects to be built in certain critical environmental areas. As an incentive to localities to develop comprehensive land-use plans, the act provides for the transfer of some of the agency's permitting authority to local governments once their plans have been developed and approved by the agency.

The Adirondack Park Act provides for far more comprehensive control by a single coordinating agency than has ever been attempted by the Federal Government. But there are precedents at the federal level also. The creation of new national parks, where already existing communities would interlock with the federal parkland, led the Congress to seek new ways of providing for a measure of federal influence over the use of private land. The first such experiment—a federal regulatory mechanism that has become known as the Cape Cod Formula—was introduced at Cape Cod National Seashore, established by the Congress in 1962.[39]

Cape Cod National Seashore: Under the Cape Cod Formula, the Park Service promulgates standards for the uses of private land within the seashore. These standards then are to be incorporated into local zoning and subdivision ordinances. Once these are adopted, the land is exempt from federal condemnation, so long as proposed development is not incompatible with these standards.

The Cape Cod Formula, with some minor modifications, has been used at a few other national park units, such as Sleeping Bear Dunes National Seashore in Michigan and Fire Island National Seashore in New York, and at several national recreation areas administered by the Forest Service, including Sawtooth and Hells Canyon in Idaho, Oregon Dunes in Oregon, and Whiskeytown-Shasta Trinity in California. While generally successful at Cape Cod, the formula often has suffered in its transference. At Fire Island, for instance, goals have not been achieved because of the apathy and antipathy of local governments and because the Park Service lacks acquisition funds necessary to condemn developments which violate Park Service use standards, particularly construction on fragile dunes.[40]

Federal initiatives using the Cape Cod formula have been applied cautiously, and with mixed success. The experience at the Sawtooth National Recreation Area, the best current example of efforts to exert a federal influence over land use, illustrates at once the hopes, opportunities, and frustrations of this approach.

Sawtooth National Recreation Area: The unique protective program in operation at Sawtooth combines Forest Service-directed control over land use and the purchase of less-than-fee interests in the land on a scale never before attempted by a federal agency.

It was the intent of Congress when it enacted the Sawtooth legislation in 1972 to preserve not only the stunning panorama of jagged peaks and snow of the Sawtooth Range and the heavily forested middle-ground slopes, but also the pastoral valley foreground where the Salmon River meanders through open grasslands.[41] If congressional hopes are fulfilled, there will be no roadside paraphernalia—billboards, motels, gas stations, or obtrusive subdivisions—to clutter up the view of the Sawtooth Range from U.S. 93, the only highway through the Sawtooth Valley.

Sawtooth National Recreation Area contains 754,000 acres. At the time of its establishment, 726,000 acres, or about 96 percent of the total, already were in federal ownership.[42] The remaining 25,400 acres were in private ownership, primarily in the valley, which provides the scenic foreground to the imposing mountains. Congress felt that most of this private land need never be brought into federal ownership—fully. Indeed, the Congress limited Forest Service acquisition to no more than 5 percent of the private acreage for direct public use, such as recreation facilities and access to lakes and streams.[43] The Sawtooth Act provides that only "such title" should be acquired as is necessary to protect the valley's natural and scenic values.

With this congressional limit on full-fee acquisition in mind, the Forest Service has embarked on an unprecedented program of scenic-easement acquisition on private lands in Sawtooth Valley and Stanley Basin. Kenneth R. Dittmer, who directs the Sawtooth lands staff, considers the Forest Service's authority to acquire scenic easements there the most significant authority in the Sawtooth Act. "This program provides the basic mechanism by which private land use is and will be guided; the scenic easement is the major tool we are using in meeting the challenge of the legislative direction. This is the first time the Forest Service has been involved to such an extent with private lands."[44] But the harsh fact is that the acquisition program is running out of money. The cost of land purchased outright and of scenic easements has far exceeded what either the Congress or the Forest Service anticipated.

By 1976, the full $19,800,000 authorized in the Sawtooth NRA Act had been appropriated. As of January 1, 1977, 910 acres had been acquired in fee and 7,403 acres in scenic easements. There were 1,595 acres of fee acquisitions and 15,042 acres of scenic easements remaining to be acquired. The Forest Service estimated that an additional $23,000,000 will be needed to complete these acquisitions and requested an increase in the authorized $19,800,000 ceiling. Further, additional monies and a larger increase in the ceiling will be needed if the exercise of mineral rights creates a serious problem.[45] In summary, the Forest Service reached its initial funding ceiling with the easement program only one-third complete.

A Forest Service request in 1976 for additional acquisition money for Sawtooth has stimulated a Forest Service study of the effectiveness of acquisition and private-land regulatory programs at Sawtooth and four other congressionally designated areas managed by the Forest Service. The effectiveness of the Sawtooth strat-

Serrated mountain crests, forested foothills, valley ranchland traversed by the Salmon River: each is an integral component of the special landscape of the Sawtooth National Recreation Area in Idaho.

egy is still far from certain. Like any experiment, it must be tested over time and judged against standards of accomplishment that constantly change. But it seems inappropriate to judge the Sawtooth experience solely against intentions and estimates of land values perhaps valid at the time of the establishment of the national recreation area, which may have been rendered obsolete by subsequent circumstances beyond the control of the Forest Service —such as the state of the national economy, demands on the federal budget, and continuing inflation and escalation of land costs.

Other Cooperative Proposals: Meanwhile, the Congress is searching for less expensive ways to protect large areas of high landscape, environmental, and recreational value. At the same time, it wants basic management responsibility to remain with state and local governments. A new concept was described in some detail by the Congressional Research Service in a 1975 report to the Senate Subcommittee on Parks and Recreation.[46] It proposed the creation of "Green Line Parks," so named with "a certain modern latitude in color choice after the historic Blue Line delineating the boundaries of the Adirondack Park." Such Green Line parks could be established, according to the CRS, from "coherent landscape areas with high recreational potential that now are (or could be) partially owned by public and quasi-public agencies, but for the most part would consist of unspoiled land still in private ownership."[47] Land would be protected through outright purchase by various levels of government, the acquisition of less-than-full-fee interests, such as scenic easements, and land-use regulation. Since much of the land would remain in private ownership, the need for expensive acquisition would be far less than for traditional parks. Some aspects of the Green Line parks concept have been incorporated in legislative proposals to establish the Santa Monica Mountains and Seashore Urban Recreation Area in California. Senator Bennett Johnson, chairman of the Senate Subcommittee on Parks and Recreation, sees this as a way of creating large recreational areas with less federal money and less need for a continuing federal presence.

Another federal initiative deserves mention: Senator Edward Kennedy has proposed a "trust area" designation for the Nantucket Islands off the southern coast of Massachusetts.[48] Under his proposal, land would remain in private ownership, but use would be controlled through an intricate arrangement involving the Federal Government, the State of Massachusetts, and three local "trust commissions" which would prepare land-use plans.

All of these efforts represent experiments with new ways of guiding land use as well as new intergovernmental relationships. All relevant governments are charged with exercising a variety of acquisition, regulatory, and incentive techniques to protect the natural integrity of a large land area of high natural value. These techniques might be applied selectively to some national forests, or specific areas of high value within national forests.

In such a situation, the Federal Government might acquire some land and purchase easements on other parcels which would remain in private ownership. A state might provide overall land-use coordination by designating the area one of critical state concern. Local units of government would continue to exercise basic land-regulatory responsibility within the framework established by the state. The Federal Government could provide planning assistance and even some funds to strengthen local planning and regulatory staffs. While the Federal Government might initiate such comprehensive land-management programs, they more appropriately should originate with the states or with local units of government.

The application of land-use controls as rigorous as those in effect in the Adirondacks, or of massive less-than-fee acquisition on the scale of Sawtooth, would not be necessary or desirable in most areas of the national forests. But in most instances, stronger coordination, firmer central guidance, and the use of techniques more flexible than full-fee acquisition are required.

LEGAL OPTIONS

In the legislation establishing Cape Cod National Seashore and Fire Island National Seashore, Congress authorized the Park Service to condemn private land within the seashore boundaries when a proposed use does not conform to standards established by the National Park Service. In the case of Sawtooth, the Forest Service was given more direct zoning authority in the absence of local regulatory action, again with condemnation as the ultimate tool to enforce compliance.

However, there are limits to the effectiveness of this technique. Condemnation is expensive and requires that Congress appropriate sufficient funds to make the prospect of condemnation an effective enforcement tool. If Congress does not appropriate adequate funds, as it failed to do in the case of Fire Island, then condemnation is an empty threat and private landowners are not discouraged from proceeding with standard-violating development. In addition, the use of condemnation as an enforcement mechanism detracts from, and may even interfere with, the agency's overall program of acquisition of lands needed for public use.

If condemnation is too costly to be relied upon as a regulatory tool, what options remain for the Forest Service to protect public lands from the adverse impacts of private land development—pollution of a national forest stream, for example? The most obvious course of action is for the Federal Government to bring action to protect the public's land just as a private landowner might halt the damaging effects of a neighbor's actions. The legal situation is far from clear, and federal agencies have been reluctant to take direct action in the absence of firm statutory authority.

Courts have ruled that the property clause of the Constitution, which provides that "Congress shall have power to dispose of and make all needful Rules and Regulations respecting the territory or other property belonging to the United States,"[49] gives the Congress the power to "protect [the federal lands] from trespass and injury."[50] In many circumstances, the prevention of injury to federal land may require some degree of restraint over activity on adjoining, nonfederal land. For example, some types of activity on adjoining land could diminish watershed values, increase water and soil runoff, or deplete an acquifer. In such situations, it can be argued that the broad power conferred on the Congress by the property clause includes the power to control activity on adjacent land as necessary to prevent deleterious effects to federal property. Cases to which this principal have been applied include an 1897 decision by the Supreme Court barring a fence on private land which blocked access to federal property and, conversely, several lower court cases in which private landowners were ordered to erect fences to keep livestock from straying onto the national forest. While the circumstances may be different in the case, for example, of soil-laden water flowing off private land onto public land, the principle remains the same.

In 1976 the Supreme Court recognized the validity of the general proposition that when the Federal Government reserves public land for special purposes, it may be necessary to control activities on adjacent or nearby land to prevent the negation of those purposes. In Cappaert v. United States,[51] the court held that when the government set aside Devil's Hole National Monument in Nevada for the protection of the rare pupfish, it assumed reserve water rights sufficient for that purpose. In order to preserve the water level required for survival of the pupfish in Devil's Hole, the court upheld an order preventing a nearby rancher from pumping groundwater on his property. While no constitutional question was involved—only whether the government had appropriated water rights before the rancher began his pumping—this case provides judicial recognition of the dependency of federal lands on the uses of neighboring private property.

Despite these and other precedents, federal agencies, including the Forest Service, generally have been reluctant to take legal action against private landowners, particularly since Congress—specifically in the creation of national recreation areas—has relied upon condemnation and acquisition of land to prevent incompatible or damaging uses.

This reluctance has raised the question of whether federal agencies require direct statutory authority from Congress to control activities on adjacent land. This situation was further confused by the June 1976 decision in the Northern District of California, in Sierra Club v. Department of Interior.[52] In this case, the Sierra Club brought suit against the Department of Interior, contending that the National Park Service had failed to fulfill its statutory responsibility to protect Redwood National Park from the effects of timber harvesting on private land on the park's periphery. The court had earlier ordered the Park Service to take such action. One course of action suggested by the Department of Interior was to propose legislation "for additional regulatory power over peripheral timber operations . . ."[53] This approach was rejected by higher administration officials.

The court relieved the Department of Interior from further obligation, declaring that "it follows that primary responsibility for protection of the park rests no longer upon Interior, but squarely upon Congress to decide whether and, if so, when, how and to what extent new legislation should be passed to provide additional regulatory powers and funds for protection of the Redwood National Park."[54]

In its reliance upon Congress to provide specific statutory authority, the Court was begging the question. Other precedents suggest, that, **even without statutory authority to directly regulate adjacent land uses, the agencies of the Federal Government have an obligation to exercise their rights to intervene in administrative hearings and to take formal legal action, if necessary, to enjoin activities on adjacent land in order to protect the value of the public's resources.** This approach would not entail direct federal regulation of land use, but rather would encourage state and local jurisdictions to exercise their police-power authority or pressure the state to enforce state environmental regulations. In such cases, the Forest Service would be asserting the same position as a private landowner threatened with damage to his property because of a neighbor's activity.

SHARING DECISIONS

If planning for entire forest regions—with all jurisdictions participating cooperatively—is to succeed, the Forest Service must be responsive to outside views about the management of the national forests. The motivation to do so should come from the conviction that integrated regional planning offers the best hope that the forests will complement other lands of the region in meeting the needs of society.

In addition to being responsive to officials of governmental units and commissions, the Forest Service should improve mechanisms for public participation in its decision-making process. The comparative autonomy that the Forest Service has traditionally exercised within its broad statutory mandates and its hierarchical, closed organizational system help explain its long-standing tendency to resist outside intrusion into what it sees as its management prerogatives. But now, as demands on the national forests increase,

The variety of produce from area farms is displayed at this roadside stand in the Nantahala National Forest in North Carolina.

the Forest Service finds itself ever more on the defensive. Conflicts and disputes arise which the Forest Service, as the arbiter of competing interests, finds nearly impossible to resolve. The goal of harmonizing uses makes denial of any single interest more difficult. A broader public involvement in national forest decision making could substantially improve the situation and ease the burden on the Forest Service.

In recent years, particularly since the advent of the National Environmental Policy Act, increasingly frequent challenges to the judgment and authority of the Forest Service have arisen. In the view of Jim Giltmier, a staff member of the Senate Agriculture Committee, the Forest Service "has traditionally been slightly contemptuous of outside interference, even on the rare occasions when it existed. While this professional paternalism worked far better than it had any right to, it finally broke down when the public began to insist on having a say in managing its lands."[55] In response, the Forest Service developed a program of "public involvement." Forest supervisors hold "listening sessions" to receive public views, and formal public hearings are held on draft plans. These formal meetings are often augmented by informal meetings with groups interested in a specific aspect of forest management. The Forest Service is even experimenting with structured role-playing exercises to encourage consensus. But officials maintain that, under their congressional mandate and as professionals in resources management, they must make the final decision, mediating use conflicts in the best interests of the forests and all of the publics involved. "The public-involvement process," states the Forest Service Manual, "in no way relieves the responsible line officers of the responsibility for making decisions."[56]

There are questions, of course, as to the identity of "the public." Is the public the residents of the immediate area who may depend on the national forests for their livelihoods? Wilderness advocates? Off-trail vehicle owners? Of course, the public is all of these and more. Representatives of all facets of the public concerned with national forest issues should have a stronger role in the decision-making process. New mechanisms are needed to better translate the concerns of forest users into Forest Service action.

Though few admit it, there is currently a general feeling of self-satisfaction among those citizens who deal on a regular basis with the Forest Service as representatives of relatively specific interests in the forests. Most are confident that they have a personal impact, and feel comfortable talking with other—even opposing—players in the game. They are content to handle day-to-day issues and problems rather than address the future aggressively. A Pennsylvanian discussing this suggested that "pressures for change are going to have to come from the cities and from people who are not now concerned or involved." And a professional New England forester observed, "If forest lands are to have a substantially different or expanded role in meeting future needs in New England, the impetus for change will have to come from outside the existing circle of the Forest Service and its interest groups."[57]

A number of questions are raised by any proposal to increase decision sharing in national forest management. Would the constituency best be represented through citizen advisory committees or commissions with strong policy-setting authority? Should decision-sharing groups focus on single forests or on geographic regions? Should the members represent the immediate forest environs, the over-all geographic region, or national interests? Or should all such categories be represented on any group established to increase public participation in national forest management?

The creation of new advisory committees or the expansion or change of existing ones must comply with the Federal Advisory Committee Act of 1972, which regulates the creation and activities of advisory committees to federal agencies.[58] The objectives of the act are to stem the proliferation of advisory committees, to make sure that they are representative of all interests, and to ensure that their activities are in the public interest.

The law limits new advisory committees to those considered essential; provides for their abolition when their purpose has been fulfilled, or after two years (unless expressly created by an act of Congress or renewed by an action of the President); limits their function to advisory activities only; and provides that only a federal employee may call and chair a committee meeting.

According to the 1975 report on federal advisory committees, only 5 of the 150 national forests have advisory committees. Three of these are in the East, for the White Mountain, Ottawa, and Superior National Forests.[59] White Mountain Forest Supervisor Paul D. Weingart says his advisory committee, which has been in existence for some years, provided him with "instant contact with my constituency" when he assumed the supervisor's position.[60] Weingart has enlarged the group to 18 members, including a state legislator, representatives of the major environmental organizations interested in the White Mountain, two lumber company executives, representatives of snowmobile and off-road vehicle associations, a ski resort owner, a representative of the association of developed recreation businesses in the area, and individuals with other expertise or interest in the forest.[61] The group meets quarterly.

One long-time member of the advisory committee, Paul Bofinger, has seen it evolve from a rather select group of citizens, each with a broad interest in the forests, to the present collection of user representatives. Bofinger, executive director of the Society for the Protection of New Hampshire Forests, says: "The advisory committee is effective. And it's important for the user groups, if for no other reason than it gives them an opportunity to sit down and talk with people they are usually shouting at."[62]

The White Mountain Advisory Committee works because of a tradition of community interest in the forest, but mostly because Weingart wants it to work and takes the committee seriously. Abiding by the provisions of the Advisory Committee Act, the committee remains a creature of the Forest Service. Weingart chairs its meetings. He and his staff prepare the agenda, although he usually discusses it in advance with committee members.[63]

Another advisory committee with a long record of success, and whose historic development and range of activity are of interest for forest-management proposals, is that for the Cape Cod National Seashore, a unit of the National Park System.[64] The Cape

Cod National Seashore Advisory Commission was created by Congress through a provision in the 1962 legislation establishing the seashore.[65] (It is now customary for national park legislation to provide for an advisory group for a specific period of time, usually 10 years from date of establishment.[66])

The commission has 10 members, one representing each of the six Cape communities and Barnstable County, two representing the state, and another the Federal Government. All were to be appointed by the Secretary of Interior, with each of the nonfederal jurisdictions to nominate two appointees, one of whom would be selected. In most instances, the Cape communities finessed the secretary by submitting a single name—thus making their own appointments, including some who had been opponents of the seashore's creation.[67]

The legislation encouraged the Park Service to consult with the commission on seashore development and acquisition. It also was to seek the "advice" of the commission before issuing any permits for industrial or commercial uses of land within the seashore, or establishing public-use areas.

While the Department of Interior at first attempted to direct the commission towards a purely advisory role, with the Park Service in control of its activities, the commission quickly established its independence. According to a recent account of the commission's

A combination general store and gas station near the Chattahoochee National Forest in Georgia.

history, "From the second meeting on, no Park Service official was ever seated with the commission at the conference table without prior invitation. Washington regulations notwithstanding, the commission's business was its own. Government was merely an invited guest."[68]

The commission won high marks from the state and local officials in a survey conducted by the commission chairman in 1975. It was described as "the only vehicle available to participate in policy formulation," a "device to maintain lines of communication," and "in the best interests of the national seashore park and the Lower Cape communities."[69] The commission is now in its second decade, its life having been extended administratively by the Park Service. Though renewed after passage of the Advisory Committee Act, the commission still functions as it did during its first decade and elects its own chairman.

Participants give much credit for its success to the Park Service: "The Park Service has, in fact, consulted regularly with the advisory commission before making any major decision relative to the seashore, just as Congress intended."[70] It also appears that strong commission leadership, the independence of the commission, and

Expensive recreation homes are arrayed across a mountain near the Pisgah National Forest in North Carolina.

the inclusion of representatives of all the Cape's jurisdictions have contributed to its success.

It is evident that there are inherent weaknesses in advisory committees at all government levels. When created by governmental agencies, they often are assigned low priority for staff assistance (and staff members frequently view work with advisory groups as an extra, unnecessary burden). Serving without pay, sometimes without reimbursement of expenses, and required to take time off from their jobs, advisory committee members often turn out to be a small group of persons with narrow interests. Appointments to advisory committees may be dispensed as political rewards, going to persons with little interest in the subject at hand. Then, too, advisory group members often feel they have little real influence over decisions and become apathetic. But such problems can and should be overcome, for the value of an advisory committee can be considerable.

The legislation approved by the Senate to resolve the legal block to clearcutting on the Monongahela authorizes the Secretary of Agriculture to "establish and consult such advisory boards as he deems necessary to secure full information and advice on the ex-

A settlement adjacent to the Ocala National Forest in Florida.

ecution of his responsibilities." Membership is to be "representative of a cross-section of groups interested in the planning for and management of the national forest system and the various types of use and enjoyment of the lands thereof."[71] But this authorization to form advisory committees is not sufficiently strong; it does not *require* their establishment, but only permits it at the discretion of the Secretary of Agriculture.

An advisory committee should be established for each major eastern national forest. (Where there are numerous small forests within a single state, it may be appropriate to appoint a single statewide advisory committee.) Each committee should be composed of representatives of all user groups, both local and national, including residents of fairly distant metropolitan areas; local residents and representatives of local economic interests; officials of all levels of government; and representatives of labor unions, consumer organizations, affected trade associations, rural advocacy associations, and other appropriate groups.

The existing advisory committees for both national forests and parks in the East, by and large, have demonstrated their worth and have proved that such committees can be useful appendages to government. The Cape Cod Advisory Commission, in the words of Park Superintendent Lawrence Hadley, "has been invaluable in helping to fit the seashore to the communities."[72] Advisory committees could do the same for national forests. The chance that they will become dominated by special interests seems to have been eliminated by the passage of the Advisory Committee Act, which also makes it clear that the land-management agency bears final responsibility. Ultimate accountability for national forest management will continue to rest with the people's elected representatives in the Congress, and this is as it should be.

In the 1960s, the Forest Service had advisory committees for many of its regions and national forests. The fact that there are now advisory committees for only 3 of the 50 eastern forests perhaps reflects the dampening effect of the strict justification requirements of the Federal Advisory Committee Act more than Forest Service disinterest. While the elimination of useless appendages of government is a laudable objective, proven avenues of meaningful public participation should not be closed. Sporadic listening sessions and public hearings cannot substitute for the continuing, formal contact between forest users and managers which advisory committees provide.

REFERENCES

Chapter V

1. Society of American Foresters, "Forest Land Use Planning: A Position of the Society of American Foresters," approved April 2, 1976.
2. William E. Shands, *The Subdivision of Virginia's Mountains* (Washington: The Conservation Foundation, 1974).
3. U.S. Forest Service, *Plan for Managing the Ocala National Forest* (Atlanta: U.S.F.S., 1972), p. 3.
4. C. Lowell Harriss, ed., *The Good Earth of America*, Publication of the American Assembly (Englewood Cliffs, N.J.: Prentice-Hall, Inc., 1974), p. 23.
5. Data supplied by the West Virginia Office of Federal-State Relations, October 1976.
6. Data supplied by the Virginia Division of Planning and Community Affairs.
7. Interview with David K. Rice, Chairman, Board of Commissioners, Warren County, Pennsylvania, July 1975.
8. Interview with Charles A. Blankenship, U.S. Forest Service, Jefferson National Forest, February 1976.
9. Arrowhead Regional Development Commission, *Impacts and Opportunities: Voyageur Planning Area* (Duluth, Minn.: Arrowhead Regional Development Commission, n.d.).
10. Interview with Wendell Doty, U.S. Forest Service, State and Private Forestry, August 1975.
11. See Robert G. Healy, *Land Use and the States* (Baltimore: Johns Hopkins University Press for Resources for the Future, 1976).
12. Phyllis Myers, *So Goes Vermont* (Washington, D.C.: The Conservation Foundation, 1974), p. 11.
13. Phyllis Myers, *Slow Start in Paradise* (Washington, D.C.: The Conservation Foundation, 1974).
14. Virginia Division of Planning and Community Affairs, *Virginia Local and Regional Planning, 1975* (Richmond: Virginia Division of Planning and Community Affairs, 1975).
15. *Land Use Planning Reports*, January 6, 1976.
16. Council on Environmental Quality, *Untaxing Open Space* (Washington, D.C.: GPO, 1976), Table I, p. 13.
17. Kenneth B. Pomeroy and John Muench, *The Challenge of Private Woodlands* (Washington: American Forestry Association, 1975), p. 37.
18. *Land Use Planning Reports*, September 13, 1976, p. 5.
19. U.S. Forest Service, *State Forester and Forest Service Accomplishments in Rural Development* (Washington: GPO, 1975), p. 9. (Hereafter referred to as *State Forester.*)
20. Ibid., p. 13.
21. U.S. Forest Service, *Forest Plan, Green Mountain National Forest* (Milwaukee: U.S.F.S., 1975), p. 21.
22. Telephone interview with Robert Butler, U.S. Forest Service, Green Mountain National Forest, August 1975.

23. U.S. Forest Service, *Land Use Plan for the Deerfield River Area* (Milwaukee: U.S.F.S., 1975), p. 33.
24. Ibid., p. 36.
25. Interview with Harold E. Anderson, forestry consultant, Arrowhead Regional Development Commission, October 1975.
26. Ibid.
27. Interview with Donald S. Girton, U.S. Forest Service, Supervisor, Wayne-Hoosier National Forests, February 1976.
28. U.S. Forest Service, *Forest Plan, Allegheny National Forest* (Milwaukee: U.S.F.S., 1975), p. 31.
29. Interview with Donald S. Girton.
30. "Kentucky Court Continues Public Land Strip Mining Case," *Land Use Planning Reports*, November 3, 1976.
31. Letter from John P. Butt, Forest Supervisor, Allegheny National Forest, August 18, 1975.
32. Ibid.
33. U.S. Forest Service, *State Forester.*
34. Adirondack Park Agency Program Report, reprinted in the Lake Placid *News*, March 4, 1976.
35. Ibid.
36. Ibid.
37. Adirondack Park Agency, *Adirondack Park Land Use and Development Plan* (Ray Brook, N.Y.: Adirondack Park Agency, 1973).
38. Ibid.
39. 16 U.S.C. 4591.
40. National Park Service, *Fire Island National Seashore Draft General Management Plan* (Washington: GPO, 1976), p. 11.
41. 16 U.S.C. 460 aa.
42. Interview with Gray F. Reynolds, Superintendent, Sawtooth National Recreation Area, February 1976.
43. Lands condemned because of violation of the Forest Service land-use standards are exempt from this 5-percent limit.
44. Interview with Kenneth R. Dittmer, Lands Staff, Sawtooth National Recreation Area, February 1976.
45. Letter from Dennis P. Grassi, Forester, Lands Staff Unit, Sawtooth National Recreation Area, December 23, 1976.
46. Environmental Policy Division, Congressional Research Office, *Green Line Parks: An Approach to Preserving Recreational Landscapes in Urban Areas* (Washington: GPO, 1975).
47. Ibid., p. 14.
48. S. 67, 94th Congress.
49. U.S. Constitution, Article 4, Sec. 3, CL 2.
50. Kleppe v. New Mexico, — U.S. —, 96 S. Ct. 2285, 2292 (1976).
51. Cappaert v. U.S. 426 U.S. 128 (1976).
52. Sierra Club v. Department of Interior-F. Supp.—(N.D. Cal. June 7, 1976).
53. Ibid.
54. Ibid.

55. Jim Giltmier, "Resources Planning Act," *Journal of Forestry*, May 1976, p. 275.
56. *Forest Service Manual*, Sec. 8212, October 1973.
57. Perry R. Hagenstein, "Perspectives on the Future Role of National Forests In New England," December 1975.
58. Federal Advisory Committee Act, PL 92-463.
59. General Services Administration, *Federal Advisory Committees, Fourth Annual Report of the President* (Washington, D.C.: GPO, 1976).
60. Interview with Paul D. Weingart, Forest Supervisor, White Mountain National Forest, August 1976.
61. Ibid.
62. Interview with Paul Bofinger, Executive Director, Society for the Protection of New Hampshire Forests, August 1976.
63. Interview with Paul Weingart, August 1976.
64. Francis P. Burling, Charles H. W. Foster, Robert F. Gibbs, "A Promise Fulfilled," May 1976.
65. 16 U.S.C. 4596.
66. One national park advisory commission—that for Fire Island National Seashore in New York—had a tempestuous existence, with members suing the National Park Service over management policy. When the advisory committee expired after its 10-year statutory life, it went unmourned by the Park Service.
67. Burling, et al., "A Promise Fulfilled," p. III-3.
68. Ibid., p. III-12.
69. Ibid., p. III-64.
70. Ibid.
71. PL 94-588, Sec. 146.
72. Interview with Lawrence Hadley, Supervisor, Cape Cod National Seashore, August 1976.

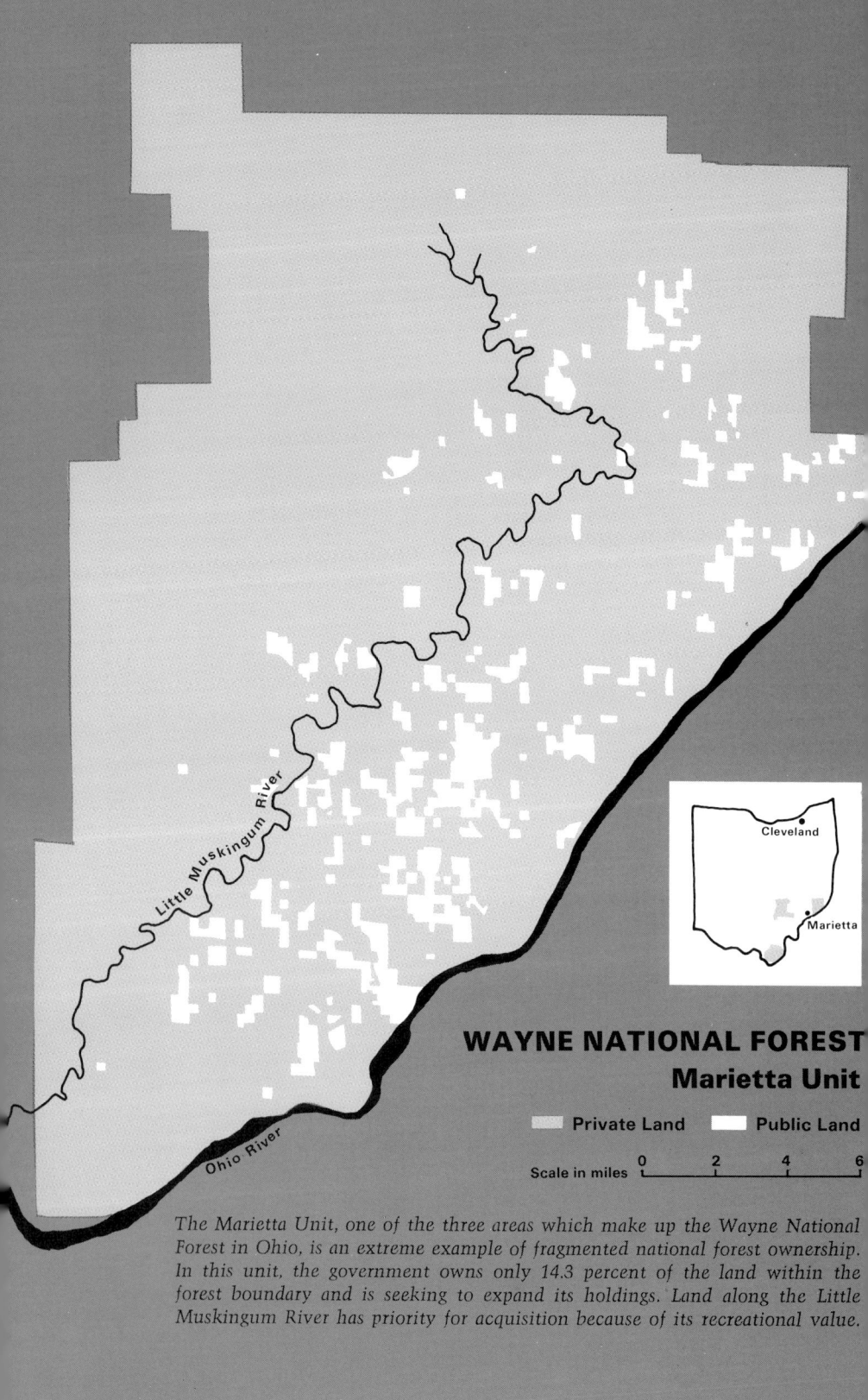

WAYNE NATIONAL FOREST Marietta Unit

The Marietta Unit, one of the three areas which make up the Wayne National Forest in Ohio, is an extreme example of fragmented national forest ownership. In this unit, the government owns only 14.3 percent of the land within the forest boundary and is seeking to expand its holdings. Land along the Little Muskingum River has priority for acquisition because of its recreational value.

CHAPTER VI

ownership patterns and acquisition

Increased cooperative efforts among the Forest Service, the public, governmental units, and landowners responsible for adjacent lands and inholdings offer probably the most comprehensive and financially realistic route toward achievement of forest objectives. The fragmented ownership pattern of this system makes such cooperation essential, while it also suggests outright purchase and exchange of land as critical management tools. In the East, where the Forest Service owns only about half the land within the established forest boundaries, acquisition is particularly attractive to land managers.

In acquiring land, national forest managers strive for an ownership pattern which protects forest resources, maximizes opportunities for forest use and enjoyment, and permits management efficiency. Federal ownership must adjust to changes in forest use, development of adjacent private land, and regional social and economic needs. These constantly changing factors make "land ownership adjustment" — as the Forest Service calls it — a never-ending process.

Clearly, though, acquisition is not a cure for all the problems of the national forests. Rather, it is but one tool used by the land manager to achieve resource-management objectives. Forest managers find acquisition particularly attractive because fee-simple, or outright, purchase brings the land under their direct control. It is often a simpler, speedier process than efforts to persuade local officials to regulate land uses that affect the forest environs. In the absence of local land-use controls in rural counties, Forest Service acquisitions are often defensive in nature — aimed at preventing developments that would damage forest resources.

Since the turn of the century, there has been considerable agreement that the Federal Government should acquire and protect large areas of forest land. The question now is: How much land and what kind of land is necessary or appropriate?

As one Forest Service official explained, policy has evolved over the years:

> We used to believe that all the land on the national forest map had to be green—in public ownership. We have long since abandoned this position. Now we ask how much green, and where.[1]

A few examples of current ownership situations illustrate the kinds of decisions about acquisition priorities and procedures that the Forest Service must face constantly:

In the Wayne National Forest, the Federal Government owns 163,345 acres[2] scattered in three units across southern Ohio. This acreage amounts to about 20 percent of the 833,288 acres within the forest boundaries. "In its present state," says a concerned Forest Service official, "it is not a viable unit" — that is, an area appropriate for management for multiple forest uses. Though the Wayne receives about two-thirds of the total annual Weeks Law allocation of the Forest Service's eastern region, at the present annual rate of acquisition it will take 75 years to acquire half of the acreage within the forest boundaries, the Forest Service's unofficial acquisition goal.[3] *Should the Forest Service continue to buy land in the Wayne, recognizing that it will take a long time to reach its acquisition goal, if indeed it ever can? Or would it be wiser to abandon its efforts there, given funding constraints, and request congressional action to permit the Wayne to be used as trading stock to build other, more promising forest units?*

In the Allegheny National Forest, the public owns the mineral rights to only 1 percent of its surface ownership.[4] Oil exploration by private owners is increasingly impinging upon the use of surface resources. (A large area in the northeastern portion of the forest is a lacework of oil exploration roads and well sites.) *Should the Forest Service embark upon an aggressive program of acquisition of subsurface rights in the Allegheny and other forests, recognizing that it will reduce funds for acquiring surface land?*

Within the boundaries of the Wayne and some other Appalachian national forests are extensive areas that have been stripped of their coal and abandoned by owners. *Should the Forest Service acquire these devastated lands and rehabilitate them — at considerable cost?*

Most forest lands accommodate several uses, including timber, recreation, and wildlife. Because of the Land and Water Conservation Fund Act, current acquisition emphasizes land of high outdoor recreation value. Funds for purchase of general forest land have been severely limited. *Should the nation put its national forest purchase dollars into general forest land — including land of high timber productivity — or into recreation land?*

While these questions oversimplify actual circumstances, they illustrate some of the national policy, funding, political, and procedural decisions which must be made.

OWNERSHIP PATTERNS

The nearly 24,000,000 acres of federally owned forest land in the East amount to only about half the land within the established national forest boundaries. On 23 of the 50 eastern national forests, the government owns less than half of the acreage within the forest proclamation boundary, the area established by Congress within which the Forest Service is authorized to acquire land.[5]

In many of the forests of the East, the anemia of public ownership is acute. Only 20 percent of the land in the Wayne National Forest boundaries is publicly owned. Several other forests are only slightly better off: In the Uwharrie National Forest in North Carolina, the Federal Government owns 21 percent of the acreage within forest boundaries; in the Hoosier National Forest in Indiana and the Holly Springs in Mississippi, the figure is 28 percent; and in the Shawnee in Illinois, 30 percent.

But even in those forests where the public owns a majority of the land, areas most attractive for development often remain in private ownership. In the mountain forests, the Federal Government typically owns the ridges and steep slopes, while valleys convert to recreational subdivisions. In the White Mountain National Forest, where the public owns 85 percent of the land, forest boundaries were drawn to omit the valleys and notches that attract development. In the Lake States forests, where ponds and lakes abound, key waterfront tracts remain in private hands.

The problem of mineral rights is even more troublesome. The Federal Government owns about two-thirds of those rights beneath its acres in the East.[6] However, in minerally rich areas, the government's ownership of minerals is as low as 1 percent (in the oil-rich Allegheny National Forest). More common are situations like that in the Wayne, where the public owns mineral rights to about 17 percent of its national forest land, and in the Monongahela, where 10 percent of the coal is in public ownership.

Small and fragmented public ownerships such as those in the Wayne, the presence of inholdings within otherwise solid blocks of public land, ownership schisms like those in the White Mountain, and private exploitation of minerals over which the Forest Service has little or no control — all of these factors create severe problems of resource protection and management efficiency. The bulldozing of land for development of a forest inholding on Massanutten Mountain in the George Washington National Forest has resulted in the pollution of a forest trout stream.[7] In the Ocala National Forest in Florida, inholdings are sites of blighting hunting camps, which also create fire danger in pine forest.[8] In the Nicolet, landowners around a lake are asking for a new access road across forest land, while protesting Forest Service plans to provide public access to the lake.[9] The exploitation of private mineral rights devastates the surface land and pollutes the streams in several Appalachian forests.

Ownership patterns create other problems, too. It is difficult to police small and isolated forest parcels against timber theft and unauthorized use. The establishment of a trail system may be impossible. Private landowners neighboring the national forest may object to timber cutting or the establishment of a campground. In the national forests of the East, thousands of acres are virtually denied to the public because private land blocks convenient access to nearby roads.

But it is easy to overemphasize the desirability of consolidated ownership in the interest of custodial convenience. Dispersed public and private forest ownership has its benefits. In the Holly Springs National Forest, where public land is scattered randomly across half a million acres of north-central Mississippi, national forest tracts provide a diverse landscape in an area where most of the private land is in farms. In the Marietta Unit of the Wayne National Forest, pastoral ridgetops and gentle valleys offer a pleasing complement to the national forest's wooded slopes. Dispersed public ownership also provides a diverse environment for wildlife, with open private land complementing dense forest habitat. While not acquired for this purpose, national forest land can break up urban sprawl and deter monotonous strip development along thoroughfares, a pleasing effect that can be seen along mountain gaps traversed by major thoroughfares in the Virginia national forests.

Without question, acquisition of forest land to protect a watershed from pollution, to guard valuable wildlife habitat or a productive timber site, to permit development of a trail system, or even to ensure the presence of open space near an expanding community is often integral to the accomplishment of national forest objectives. Decisions on whether or not to exchange land with a private landowner, acquire a particular forest tract, or in general increase the public ownership within a forest should transcend the circumstances of the individual forest and involve regional and national resource objectives. Such factors as the availability of other public land in the region, local economic and development objectives, cropland preservation, and the quality of often-distant urban environments must also be taken into account.

HOW MUCH LAND TO ACQUIRE?

Then, how much public forest land does the nation need in the East? The answer, not unexpectedly, has varied over the years according to the circumstances of the moment and perceptions of national needs.

In 1912, when acquisitions under the Weeks Law first came under consideration, Forester Henry S. Graves took a deliberate

A farm in the valley, national forest on the ridge behind.

approach to acquisition. In his *Report of the Forester* for that year, Graves wrote:

> It is unnecessary to acquire all of the land within the designated areas. In some, probably not more than half the acreage should ever be recommended for purchase. Many valleys of fertile agricultural lands are included which it would not be wise to acquire. It will probably not be necessary to acquire all of the mountainous nonagricultural land with the areas. There is every reason to believe that the purposes of the government may be fully subserved by the acquisition of compact bodies each containing from 25,000 to 100,000 acres well suited for protection, administration and use.[10]

Graves added that these forest units could be 10 or 25 miles apart. He apparently saw no problems in leaving large areas within forest boundaries in private ownership, although he did acknowledge that ". . . it will doubtless be practicable to cooperate with surrounding private owners in fire protection and conservative lumbering."[11]

But in 1934, as the nation began its most ambitious period of national forest establishment, the authors of the massive *A National Plan for American Forestry* (known as the Copeland Report after its congressional sponsor) had expansive ideas:

> Public ownership is the only remaining alternative for chief reliance in meeting national requirements [for forest lands]. To be thoroughly effective, however, public ownership would require a program of such proportions that it would rank among the largest that have ever been undertaken by the American people. But under normal conditions the American people have never allowed themselves to be frightened out of a necessary program by mere size and cost.[12]

The most recent official statement on the subject is the 1970 report of the Public Land Law Review Commission, *One Third of the Nation's Land*. The commission, established to conduct a comprehensive review of public land law and policy, decided that ". . . a major enlargement of national forests . . . through the acquisition of private land would not, in our view, further any contemporary national purpose."[13] The commission added: "We would not preclude further acquisition of lands by the Forest Service . . . but we recommend that Congress limit the purposes of such acquition to inholdings, boundary and other land tenure adjustments to facilitate better management of the units already established and to acquire access to these properties."[14] While endorsing acquisition for management efficiency, the commission saw no need to "block up" fragmented holdings simply for the sake of consolidation. And any large-scale acquisition should be specifically authorized by Congress, the commission declared.

Currently, representatives of diverse forest interests, whether supporting or opposing an increase in the scope and pace of national forest acquisition, express uneasiness about the present situation, in which congressional and Forest Service acquisition policy is unclear. Many seem wary of a vastly accelerated acquisition program. Local officials oppose large-scale forest purchase. Some have claimed that the removal of additional land from the tax rolls constitutes financial harassment of many poor rural counties, although recent legislation providing for a per-acre sum in lieu of taxes (see pages 225-231) may mitigate this problem. Some oppose forest purchase because it limits areas available for private development or for the expansion of towns. A county commissioner in the Allegheny National Forest was bitter about the extension of the forest's boundaries across the Allegheny River, even though it was approved by a majority of the county board. "We thought we had an agreement with them not to cross the river. If they keep going, we won't have any county left." [15]

Approximately one-third of the land in Warren County is publicly owned. Originally, the boundary of the Allegheny National Forest was on the east bank of the Allegheny River. Now, the Forest Service is well over on the west side. Any land the Forest Service buys must be cleared through the Board of Commissioners. We object to nothing they wish to buy within the present forest boundaries, but we highly object to anything they wish to buy west of the river. So far, the only land they've got west of the river has been by gift: the Western Pennsylvania Conservancy bought about 5,000 acres on that side and gave it to the Allegheny National Forest.

From a pure politician's standpoint, whose main objective is to create and maintain enough of a tax base to keep the county government going, the more land that goes into the Allegheny National Forest, the less income the county has.

DAVID K. RICE, Chairman, Board of Commissioners
Warren, Pennsylvania

Timber interests also speak against massive federal acquisition, saying that the transfer of forest land from private to federal ownership places even more timber under what they consider unnecessarily tough federal environmental standards for harvesting. And residents of the Mount Rogers area of Virginia have protested the condemnation of land and subsequent removal of mountain families as land was acquired for the Mount Rogers National Recreation Area. A similar situation earlier had engendered powerful opposition in the Monongahela's Spruce Knob-Seneca Rocks National Recreation Area, where Congress consequently imposed limits on the Forest Service's use of condemnation.

Yet there are also arguments in favor of an increased purchase level for national forests. The designation in 1966 of the Redbird Purchase Unit in Kentucky, the most recent large national forest unit to be established, attracted widespread support at the local, state, and congressional level when national forest designation was seen as a way of facilitating the restoration of land that had been strip mined. Some local governments have supported the Forest Service acquisition of inholdings when they awakened to the cost of servicing them if developed. Despite substantial state opposition to national forest acquisition, in May 1975 the town fathers of Stoneham, Maine, voted 8-0 in favor of a 1,400-acre addition to the White Mountain National Forest which was facilitated by The Nature Conservancy.[16] In 1972 the Top of Alabama Regional Conference of Governments asked the Forest Service to examine the feasibility of creating a new national forest in northern Alabama "for the development of a recreational base" in the region.[17]

The United States will continue to require large amounts of forest land—for environmental enhancement, recreation, and commodity production. But it is impossible today to say just how much forest land in the East should be in public ownership by the end of the century and how much of that should be in national forests. To a large degree, future needs for some forest activities and commodities and the supply available will be determined by the influence of market forces upon the private sector. "How do we know what national forest land we need until we see what the private sector will provide free of charge?" one planner in New England asked. If the price of timber increases, for example, the private sector can be expected to retain more land in timber production.

Virtually all interest groups seem agreed that the need will continue for federal ownership of national forests. To meet this need, new land should be purchased to achieve specific forest objectives and regional and national needs. But not all of the forest land and associated activities need be supplied by the public sector. In those areas where land should be publicly owned, the Federal Government need not be the owner or manager.

GOALS FOR FUTURE ACQUISITION

In the future, national forest acquisition should be targeted to forests and regions that meet broad criteria relating to special resource and environmental features of the forest (scenic areas, critical wildlife habitat, and unusual timber land), as well as the management capability of nonfederal agencies; and the availability of other public land within the region. Areas considered for national forest purchase, either large new forest units or sizable additions to present forests, should present an unusual combination of environmental and resource values of multi-state, regional, or national significance. Acquisitions should either enhance the ability of the national forest to supply high-quality resources or should protect the distinctive character of the public forest. An example of the former is acquisition of wildlife habitat in large, solid blocks in order to facilitate species restoration. An example of the latter is purchase of inholdings about to be developed for incompatible uses. Within a proposed acquisition region, the Federal Government should be the only public authority capable of and willing to accrue and manage forest land on the scale required.

There is private land in New England that provides virtually as many— if somewhat different—public benefits as the national forests do, including recreation. You don't necessarily have to condemn or acquire the land or hit the private landowner over the head. If the land is managed so its productivity doesn't decline and recreation use and other values are considered, you have the same as national forest land, no matter who owns it. The question is whether you can make the controls over private land permanent. The neat thing about the national forest is that you know it is always going to be there; with private land, there is no assurance that it won't be subdivided. The private landowner will say, "I want to keep it this way—in forest." The public has to reply, "Give us some assurance." There will have to be compensation, because the public is limiting the landowner's future options.

The problem is that we see the national forests in the traditional concept. We really should start by asking, "Why do we want national forests?" If it is for recreation, to insure a supply of timber, for wildlife

and so on, then it really isn't necessary for the Federal Government to own every acre. There are ways to get the private sector to do the same thing.
PAUL O. BOFINGER, Executive Director
Society for the Protection of New Hampshire Forests
Concord, New Hampshire

In assessing a proposed unit's resource and environmental values, special attention should be paid to its potential for providing or protecting valuable wildlife habitat, particularly for endangered species. This is currently a low priority in Forest Service acquisition policy compared to recreational or general forest use.

It is tempting to recommend that the Federal Government divest itself of those small forests—or even units within forests—where public ownership is meager and fragmented and the possibility of future acquisition low. Such forests—the Wayne in Ohio, the Oconee in Georgia, and the Okmulgee Unit of the Talladega in Alabama are examples—could then be used as trading stock to consolidate the ownership in national forests elsewhere or perhaps be transferred to state agencies, relieving the Federal Government of management responsibility. Divestiture of such small, fragmented holdings might result in a more vigorous system of national forests in the East by directing acquisition funds and manpower to those larger forests of high public multi-use benefit.

But it seems more challenging and ultimately useful to turn these into demonstration forests, illustrating to private landowners timber management sensitive to water quality and wildlife values, for example. Such forests could serve as catalysts for sound resources management within a larger area. To implement this goal effectively, the Forest Service should forge much stronger links with local landowners, and state foresters should direct their assistance programs to areas adjacent to the national forests.

National forest land should not be sold outright to private parties, for a breach in the present legal barrier to national forest sale would leave the Forest Service vulnerable to irresistible political pressure to sell choice areas for future development.

ACQUISITION OF MINERAL RIGHTS

Most casual forest users are concerned only with the forest surface and its recreational and scenic values, or timber and wildlife. But another dimension of the forest—the subsurface estate—profoundly affects surface uses. The government has little or no control over about one-third of this area under its eastern national forest holdings (see pages 58-60), to which others own the mineral rights. The public's surface rights are subordinate to the subsurface, and the mineral owner has nearly unfettered access to the underground resources.

Let's talk about surface disturbance. Oil wells are put into production on a five-spot area. Now five-spot means there are four water-injection wells and one oil-producing well in the center, and all of these wells have to be fed by access roads, water pipe lines, oil transmission lines, and electrical lines. We're speaking of well-jacks in operation—each set over a well— and they're set with time-clocks, so that they come on at various intervals during the day. On the Allegheny National Forest, a lot of these wells are adjacent to our recreation areas. They could result in injury to kids who might be climbing around on them when they automatically start up. Obviously, they can be quite a liability.

A lot of the activities associated with the oil interfere with our forest management practices. We have thousands of these wells on the national forest. When they tear up a piece of ground, it's usually for a road access. The tracks vehicles wear in the road can be two- and three-feet deep. And the only way that they can get a truck to a well is by using a D-6 bulldozer to pull it in. We can't get them to come back in and rehabilitate the road system, and that's a big problem.

The Forest Service can only demand rehabilitation of surface lands to the degree that we can exercise what we call the Secretary of Agriculture's rules and regulations. There are four different sets promulgated over the years, beginning in 1911. Which rules apply depends on when the subsurface rights were acquired, but the earlier rules don't have enough teeth to enable us to make the subsurface owner rehabilitate the surface. In many instances, the outstanding rights of the subsurface owner include ingress, egress, and regress, and they have the right to cut the timber for the oil operation.

We're taking a critical look at some of the areas that we want to develop, and instead of buying like we have in the past—just the surface—we're buying everything in fee, so that we have better control of these outstanding interests, especially around highly developed areas like the Allegheny Reservoir. We have paid as much as $740 an acre for the land we bought around the dam, and that was in a rather old, established, but half-way depleted oil field. And although these people can apply for leases through the Department of the Interior, most of these leases are not granted except to a proven prudent operator or to an area which we feel can justify recovery of the oil without irreparable damage to the surface.

ROBERT L. FIELDS, Lands Staff Officer, Allegheny National Forest
Warren, Pennsylvania

Conflicts are not inevitable in all cases of divided surface and subsurface ownership. In many instances, the minerals are now of low value; though they may be in demand sometime in the future, extraction is not imminent. In some forest areas, oil and gas can be extracted in ways that cause no serious or permanent environmental problems. In these situations, mineral development can be controlled through regulation, and outright purchase of mineral rights is unnecessary. But in some forests, the mineral problem is acute. The extensive oil extraction on the Allegheny National For-

Trucks and equipment at a drilling site in oil fields near the Allegheny National Forest in Pennsylvania.

Through pressure lines like these awaiting use beside a drilling rig, sand and liquid are pumped under pressure to fracture the oil-bearing sand beneath the surface to increase the oil flow.

est and coal mining on the Wayne and Monongahela are probably the best examples.

The cost of acquiring all outstanding mineral rights beneath the national forest subsurface is incalculable, but certainly would be immense and prohibitive. The Bureau of Mines has estimated the value of recoverable coal beneath just two scenic areas of the Monongahela National Forest at $191,000,000.[18]

An alternative to purchase is government regulation of the manner in which the minerals are extracted, including strict environ-

Workers at a modern rotary drilling rig near the Allegheny.

With drilling rig removed, a well jack pumps "Penn grade" oil, valued for its lubricative properties, from beneath the Allegheny National Forest.

mental controls and requirements for restoration of the land. Regulation of privately owned minerals, including those beneath the national forest surface, has been left to the states. States also bear primary responsibility for control of water pollution and erosion which may result from mining.

However, the strength of regulation and enforcement varies widely from state to state, and some states have been far more effective than others in controlling the environmental impacts associated with mineral extraction. This is particularly true of surface mining of coal.

The Forest Service should develop its own action plan for minerals management, particularly for forest areas where the surface resources are of high value. The action plan would involve:

• **An assessment of the potential for economic recovery of all types of minerals,** including (but not limited to) coal, oil, and gas. Factors to be considered would include the current and projected demand for a particular mineral, its quantity, quality, and accessibility, and the attitude of the mineral rights owners toward minerals development (some might be willing to extract the minerals in nondamaging ways).

• **An assessment of the potential effectiveness of available state and federal regulatory authority to control the environmental impacts of extraction.**

• **A plan for acquisition if the regulatory authority is considered to be inadequate to protect the resource.** Such a plan would identify lands to be acquired in order of priority, based on factors such as value of the surface resources, potential for economic recovery of minerals, effectiveness of existing regulatory programs, legal rights of mineral owners, and imminence of mineral activity. The plan also should weigh the benefits of buying the mineral rights to a given area against the benefits of using the funds to acquire additional lands elsewhere.

Strong congressional interest in federal legislation to control surface mining of coal offers the Forest Service an unusual opportunity to begin to implement a minerals management plan and deal with an immediate minerals problem. (This does not, of course, address the problems of extraction of other minerals on the national forests, or the deep mining of coal.) Because of the disparity among state regulatory programs related to coal and the fact that the efficacy of some state programs is questionable, federal legislation to establish uniform standards for all surface coal mining operations has been gaining support. Congress, in fact, twice passed legislation, in 1974 and 1975, to regulate surface mining of coal, only to see it vetoed by President Ford. An attempt by the House to override the last presidential veto failed by only three votes. Representative Morris K. Udall, who became chairman of the House Interior Committee in January 1977, declared before the 95th Congress opened that "strip-mining legislation is first on the agenda" for the new Congress. "This nation," Representative Udall declared, "needs a good, tough, sensible strip-mining bill." [19]

Surface mining for coal is an immediate problem on a number of Appalachian forests. What might federal legislation to regulate such mining mean for surface mining for coal on these forests? An analysis of the 1975 legislation (H.R. 13950—the most comprehensive of the several versions of the legislation) provides some indication of the future federal regulatory approach to all coal surface mining and its effectiveness in the control of privately owned minerals below the national forest surface.

First, the House bill would have established environmental performance standards for control of water pollution, soil erosion, and the restoration of the surface-mined site to a condition capable of supporting the same or better uses. Each state would be required to develop a program which, at a minimum, would meet the federal requirements for surface-mining regulation and enforcement. A key element of state plans would be the identification and designation of areas "unsuitable for surface coal mining." If a state failed to develop an adequate program, the Federal Government would be empowered to promulgate regulations for the state. Thus the federally designed minimum performance standards would be imposed nationwide. This should result in improved surface coal mining regulation and enforcement generally.

The proposed legislation contained specific provisions for the control of surface coal mining on federal lands, requiring the Secretary of Interior to develop controls at least as rigorous as those required of the states. Parallel with state designation of "unsuitable" areas, the Secretary would also be required to identify federal lands "unsuitable for surface mining" and strictly limit — or prohibit — surface mining there.

However, the law specifically notes that any controls over surface mining on federal land are "subject to valid existing rights."

The federal lands provisions of the bill would apply congressionally imposed minimum standards for surface coal mining on all categories of federal land *where the minerals are in federal ownership*. It does not authorize special treatment—including the imposition of standards higher than those of the state—where the minerals are in private ownership, as many are beneath the eastern forests, and where "valid existing rights" have been established.

To be sure, passage of federal surface coal mining legislation would require that, at a minimum, any coal development on the eastern national forests meet either the environmental performance standards which are applicable nationwide, or the state requirements if they are stronger than the minimum controls required by the federal legislation. In some eastern forest situations, where no extraordinary scenic features or superior surface resources are involved, these controls should be sufficient. However, there undoubtedly will be situations where, because of the high value of surface resources or scenic features, the Forest Service will want to impose tighter controls, possibly prohibiting any type of surface coal mining. "Valid existing rights" could, under the terms of the law, prohibit such action, and would require recourse to the minerals management strategy suggested earlier.

However, there are uncertainties as to the exact interpretation of "valid existing rights." The legislative history of a 1975 strip-mining bill (H.R. 25) indicates that the phrase "subject to valid existing rights" is intended to mean "subject to previous state court interpretations of valid existing rights."[20] Court interpretations vary among the states. In West Virginia, for example, a state court has held that if strip mining was not practiced in the county at the time the minerals were reserved (when the surface was acquired by the Forest Service), then the deed must specifically mention surface mining as a possible method of recovery before a "valid existing right" exists. This court ruling has halted some surface mining on the Monongahela National Forest. However, courts in other states may not come to the same conclusion.

There is, however, a way in which the valid existing rights question might be addressed in the context of the proposed legislation. Under the "unsuitable lands" provision of the legislation, and in conjunction with each state's designation of unsuitable nonfederal lands,[21] the Forest Service could — and should — establish strict standards (including outright prohibition in sensitive areas) for surface mining on all national forest land regardless of who owns the mineral rights, and leave to the courts the determination of whether or not mineral owners possess "valid existing rights." In cases in which a mineral owner proposed to use methods not complying with the Forest Service standards—or where he was barred from extracting minerals—he would have to argue in court that the regulations infringed on his rights. If, as could be the case, the court were to rule that the restrictions were indeed unreasonable, the court would determine the dollar amount by which enforcement of the standards would result in damage to the owner. This would provide a basis for the government's payment to the owner for the added cost of extracting minerals in the prescribed manner —or not at all. (The legislation should provide the Forest Service with the option of relaxing its regulation if the court-determined cost of damages were prohibitive.)

This does not mean that the Forest Service should ignore existing rights. Indeed the presence of valid existing rights should be a dominant factor in determining which mineral rights to acquire outright. Because the Forest Service will be seeking ways to get the most for its limited acquisition funds, a prudent strategy would be to acquire initially the mineral rights only to those lands which it judges to be totally unsuitable for mineral extraction—and where there is a high probability of imminent extraction—but where the owner's "valid existing rights" to develop would unquestionably be upheld in court. In these cases, it would be more cost-effective for the Forest Service to buy the mineral rights outright than to incur senseless litigation costs.

With or without a federal surface mining law, and in the case of any mineral extraction by any method, the Forest Service should make it very clear that it will aggressively exert its legal rights, as a surface owner or an adjoining property owner, to whatever protection the courts can afford it. In many cases, the threat of a damage suit, coupled with the probability of strict enforcement of federal and state pollution laws, would encourage mineral owners with extraction plans to agree with the Forest Service on mining methods and land restoration.

There may be instances, however, where conventional regulation will not adequately prevent impairment of the surface resources. For example, even if the oil beneath the Allegheny National Forest were to be extracted in ways that protected the purity of streams, the extensive road network favored by the oil companies would continue to affect other uses of the forest adversely. In such circumstances, the Forest Service should test a program of mineral rights acquisition and subsequent leasing (when in the national interest), under strong controls. The Forest Service as owner then could impose more rigorous regulations than the conventional environmental protection measures.

In Chapter III, reclamation and rehabilitation of land devastated by strip mining was suggested as a new mission for the Forest Service. Although there are Forest Service programs aimed at restoring strip-mined lands in federal ownership, new areas usually are avoided if they would require substantial investments of funds for reclamation or might make the Forest Service responsible for eliminating acid pollution of water courses.

In view of the vast areas needing attention and the funding limits and priorities for acquiring recreational and general forest land, **in acquiring devastated lands the Forest Service should begin with areas within national forest boundaries, particularly those which affect the use of nearby national forest lands.**

ALTERNATIVES TO FULL-FEE ACQUISITION

Historically, federal land management agencies such as the Forest Service and National Park Service have been antipathetic to anything other than the acquisition of full title to land. They argue that acquisition of less-than-fee interests, such as development rights and scenic easements, often costs nearly as much as full acquisition, and that enforcement of restrictions on the use of private land proves a heavy administrative burden.

However, this position is now being re-examined, largely because of congressional action. In establishing national recreation areas and wild and scenic rivers, the Congress has limited the amount of land which can be purchased outright, or acquired in fee. In the case of Sawtooth National Recreation Area in Idaho, for example, where Forest Service full-fee acquisition is limited by statute, the Forest Service plans to place more than 19,000 acres of Sawtooth Valley ranchland under scenic easement—at an estimated cost of $42,800,000.[22] (See pages 182-183.) The Forest Service, again under congressional mandate, is also acquiring easements along some wild and scenic rivers. However, the purchase of easements and other rights to the land remains little used in national forests generally.

In some cases, there is clearly no substitute for conventional acquisition approaches. But less-than-fee arrangements may be en-

If we are going to see significant national forest expansion—or simply forest protection—we are going to have to use techniques beyond the traditional full-fee acquisition—acquisition of easements, special land rights, "recreation" rights, or leasing timber, for example. The Forest Service is not going to get much more acquisition money out of Congress—not with competing demands by the Park Service, states, and cities. We must get the most protection for the dollar. That will be through purchase of less-than-full rights to the land and through the use of land-use controls, such as zoning.
PAUL O. BOFINGER, Executive Director
Society for the Protection of New Hampshire Forests
Concord, New Hampshire

tirely adequate to control access, buffer natural areas and areas of high recreational development, and ensure reasonable control while permitting compatible development. The purchase of less-than-fee interests and such related techniques as purchase and lease-back might be used advantageously to deal with forest inholdings, to preserve scenic vistas, or to perpetuate nonforest agricultural use of tracts adjacent to public forest land. In Florida, the state has leased hunting rights to private land. This might be another way to provide a financial incentive to keep land in an undeveloped state.

The Forest Service should experiment with the acquisition of less-than-fee interests in key inholdings in eastern national forests. While the price of these rights will be high—some perhaps nearly equaling full-fee purchase—and enforcement problems will remain, substantial benefits should result. Resistance to Forest Service acquisition should be reduced, since the land will remain on the tax rolls, although perhaps at reduced assessed values. Residents will retain their homes, eliminating the need to pay relocation costs. Certain nonforest agricultural lands could remain in productive use, with their scenic values protected.

FUNDING FOR EASTERN NATIONAL FOREST ACQUISITION

Two congressional appropriations provide funds for national forest acquisition: a general forest appropriation (known as Weeks Law funds) and appropriations from the Land and Water Conservation Fund, a fund amounting to $300,000,000 annually at the end of 1976, which is divided among federal agencies and state and local units of government for the purchase of recreational open space.

Largely because of funding imperatives and the LWCF mandate, national forest purchases in the last decade have emphasized recreation land, mostly within congressionally established national recreation areas and along rivers in the national wild and scenic river system. The modest Weeks Law appropriations are allocated to national forests where public ownership is sparse and to acquisition of key tracts in a few other forests.

The Weeks Law's provisions authorizing land acquisition to protect the flow of navigable streams and for timber production permit the purchase of general forest land. While there are no regional restrictions on the use of Weeks Law appropriations, most of the money is spent in the eastern national forests. Weeks Law funding has roller coastered from its initial levels of $2,000,000 a year to a peak of $33,000,000 (for a two-year period) during the Depression. The funding nadir for these appropriations came during World War II, with no money in 1944, and in the 1950's, when appropriations often amounted to less than $100,000 a year. Since 1969, Weeks Law funding has stabilized at around $1,500,000. (The President requested $1,700,000 for fiscal 1977.)

At this funding level, the Weeks Law has limited significance for eastern national forest acquisition. With $1,500,000 appropriated for Weeks Law acquisition in fiscal 1976, more than $1,000,000 went to purchase land in just three forests: the Daniel Boone, for its Redbird Purchase Unit in Kentucky ($530,000); the Ozark National Forest in Arkansas ($230,000); and the Wayne National Forest in Ohio ($218,039).[23] For the other 47 eastern national forests, hardly more than small change remained.

But the need for Weeks Law funding is substantial. In 1973, the supervisors of national forests in Missouri, Illinois, Indiana, and Ohio estimated their aggregate need for general forest purchase money at $235,000,000, with $175,000,000 of that in Weeks Law appropriations.[24]

Since its inception in 1965, the LWCF has provided more than $200,000,000 for national forest purchase, nearly twice the amount appropriated in the entire 65 years of the Weeks Law. LWCF money dominates forest purchase in the East. In 1976, the Chequamegon in Wisconsin was to receive $30,000 in Weeks Law money (a substantial amount under the fiscal circumstances) and $98,420 from the LWCF; the Green Mountain National Forest received no

Weeks Law money, but $910,637 from the LWCF; the Monongahela, $20,000 from Weeks Law appropriations and $1,300,000 from the LWCF (which went to purchase land in the Spruce Knob-Seneca Rocks National Recreation Area).[25] Substantial LWCF funds were distributed widely among the southern forests. The Jefferson National Forest in Virginia received $1,800,000 for the Mount Rogers National Recreation Area; the Chatahoochee, $705,000; the Pisgah and Nantahala, a total of $640,000 for Appalachian Trail acquisition; and the Cherokee National Forest, $600,000. The George Washington National Forest and national forests in Alabama and Florida also received significant LWCF money.[26]

While the Land and Water Conservation Fund has given the Forest Service a much-needed infusion of funds to meet recreation demands, its impact is diluted because the lands acquired are so costly. The price of forest land, particularly sites of high recreation value, is soaring. In 1970, the average per-acre price for land in the eastern national forests was $88.57; in 1975, $268.41.[27] National forest officials in Missouri report that "land prices throughout the forest have risen sharply over past years . . . land costs appear to be inflating at a rate of 10 percent a year."[28] In some instances, land prices have risen completely beyond the financial reach of the Forest Service.

Land prices around the Ocala have increased three- to ten-fold since 1971, to $1,000, $1,500, even $3,000 per acre. The 1971 price for open pasture land was about $300 per acre. Now a good five-acre tract would begin at $1,500 per acre. See this subdivision surrounded by national land? Four or five years ago, it was an old farm. It was sold and subdivided. Now it's a mini-community—and totally out of reach of Forest Service acquisition.
WALTER A. GUERRERO, District Ranger, Ocala National Forest
Ocala, Florida

At present Weeks Law and LWCF funding levels, acquisition within the existing forests will continue to move at a frustratingly slow pace. In fiscal year 1975, the Forest Service completed the purchase of only 33,537 acres in the East, with just over $9,000,000 in both Weeks Law and LWCF money.[29] Of this total, just under $1,000,000 in Weeks Law money purchased 8,606 acres.

Weeks Law money is of special importance because it can be used to buy land for many purposes (unlike LWCF funds which are restricted to recreation land), including to protect public forest holdings from the impacts of development. Clearly, even moderate and deliberate acquisition of eastern forest land will require a substantial increase in funds for general forest acquisition. **The Congress should appropriate not less than $10,000,000 per year for forest acquisition under the Weeks Law, and this annual funding level should be assured for a minimum of five years.** A guaranteed funding level for a specified period is essential in order to develop a coherent long-range plan for forest acquisition.

The $10,000,000 annual appropriation recommended for general forest acquisition is not an extravagant sum—about the cost of eight miles of interstate highway in a mountain region. It would permit the Forest Service to accelerate its acquisition program in those forests, such as the Wayne in Ohio, where it is attempting to build the public's ownership, as well as to capitalize on opportunities to buy key inholdings and adjacent land parcels in other forests. Assuming an average cost of $150 per acre (slightly more than the average 1976 price for general forest land), $10,000,000 per year would buy about 65,000 acres. This would amount to about 1300 acres in each of the eastern forests, although the funds certainly should not be allocated on such an arbitrary basis.

The Land and Water Conservation Fund will continue to be the principal instrument for acquisition of prime outdoor recreational land to satisfy the nation's needs. Funding should be increased for the acquisition of recreation land of all kinds in all areas of the nation. Delays in purchase of open space simply add to the cost and decrease the amount of land that can be acquired with available funds. Additionally, prime land is often lost to development.

Fortunately, the Congress has acted to increase the LWCF authorization from its present level of $300,000,000 per year to $600,000,000 in 1978, $750,000,000 in 1979, and $900,000,000 through fiscal 1989. This will result in a substantial amount of additional money for outdoor recreational land acquisition, including recreation land in national forests.

In its 1972 report on the National Park System, *National Parks for the Future,* The Conservation Foundation concluded that the responsibility for providing park and recreation programs within cities should remain with local jurisdictions, but recommended that future national park acquisition priority be given to "natural lands proximate to large urban concentrations."[30] Just as the provision of urban parks and recreation was deemed inappropriate for the Park Service, it would also be inappropriate for the Forest Service to embark on the acquisition of land close to cities primarily to satisfy urban recreation demands. Nor should the Forest Service build recreational facilities in the forests that replicate urban types of recreation, such as basketball courts and swimming pools. The Forest Service, commendably, has concentrated on recreation suited to the forest environment, and should continue to do so.

This does not mean, however, that the eastern forests should not help meet the needs of urban dwellers for extensive open space and natural areas reasonably accessible to the cities. In recent years, the national forests, particularly in the East, have increasingly helped satisfy this demand. In many instances, national forests relieve the pressure on the more popular national parks, as the Thomas Jefferson and George Washington National Forests in Virginia do for Shenandoah National Park.

Under provisions of the Land and Water Conservation Fund Act, at least 40 percent of the fund is to be shared by federal land-management agencies; the remainder goes to state and local governments. Since the fund's inception in 1965, Congress has allocated a majority of the federal portion to the Park Service, with the rationale that a lot of money was needed quickly to acquire land in the many new national park system units (most of them in the East) established by Congress over the last decade. Thus, from 1965 through fiscal 1977, the National Park Service had been allocated $828,500,000 from the LWCF, the Forest Service $304,700,000, the Fish and Wildlife Service $58,200,000, and the Bureau of Land Management $7,000,000.[31]

The priority given to acquiring national park lands with LWCF monies seems entirely appropriate. Congressional action creating new parks stems from broad public demand for the preservation of exceptional landscapes and recreational areas of national significance. The need for rapid infusion of funds for national park purchase is unarguable. Lands planned for addition to the new national parks—particularly those near cities—are escalating in price at phenomenal rates. Often the prospective loss to private development of a unique or spectacular scenic feature demands high acquisition priority. The publicity attending congressional action creating a national park promptly generates visitors who must be accommodated.

However, **there are areas on the eastern national forests that are accessible to major metropolitan areas and also help satisfy much the same demands for recreational opportunities and extensive** open space as the national parks. Such areas should receive equal consideration with other federal recreation lands in the allocation of LWCF monies.

With respect to public relations and the Forest Service acquisition program, it is clear to me that they have done a far superior job in the western United States than in the East. They are imaginative and innovative in that part of the country and have managed to compete very effectively with the Park Service and other extensive land-recreation activities of states. Why the great difference, I don't know. The Forest Service is less visible here, and not just for city people.

The much more urbanized tradition of the east may dictate a very different acquisition strategy than is necessary in the west, and might involve metropolitan greenbelt systems, river systems, linear forest and water-protection zones not on the agenda of other federal agencies, and watersheds serving metropolitan communities. In protection as well as service, there is tremendous potential for the Forest Service in the East.

I think the Forest Service should acquire more land near metropolitan areas, where the people can get to it. Land is still cheap, I believe, and no matter how expensive it gets, to put land in public trust is the best buy a public agency can make. It is an extraordinarily efficient way to spend the public's money. Land acquisition is a good investment—a good national buy—because the land increases in value.

BETTE WOODY, Boston University
Former Massachusetts Commissioner of Environmental Management
Boston, Massachusetts

The Conservation Foundation's proposed acquisition program is modest—too modest, some will say. But it is a pragmatic program which can be achieved if there is even minimal agreement among forest interests on objectives for eastern forest acquisition. Beyond this program, even the possibility of creating entirely new forests in the East should not be ruled out. A New England conservationist, while not disagreeing with a moderate approach to acquisition, commented, "Who knows what we might do over the next 25 years if resource needs demanded it, the political situation were right, and funds were available?"

PROCEDURAL PROBLEMS

On a procedural level, much could be done to make it more attractive for private parties to deal with the Forest Service in matters of land acquisition. For example, only recently have recreation and natural environmental values begun to take their place beside timber and other commodity values in the eyes of Forest Service appraisers. In the Spruce Knob-Seneca Rocks National Recreation Area of the Monongahela National Forest, the appraisers' failure to recognize recreation, aesthetic, and other noncommodity values meant that landowners were offered very little for their property. Court awards in some instances amounted to as much as 10 times the original Forest Service offer.[32] It is small wonder that relations between the Forest Service and private landowners often are strained when condemnation and acquisition are at issue. In West Virginia, complaints from landowners led Senator Robert Byrd to attach a rider to an appropriation bill which permits condemnation by a public agency only after land has been openly advertised for sale on the market. This gives a landowner the opportunity to realize a higher price for his property based upon what a private party might be willing to pay for the land's amenity value. Unfortunately, the provision encourages landowners to subdivide their property, and thus increase the cost to the Federal Government.

Private agencies, such as The Nature Conservancy and the Western Pennsylvania Conservancy, have helped facilitate the acquisition process. They have the flexibility and capacity to negotiate with willing sellers, and upon agreement, to purchase and hold property until the Forest Service can acquire it. However, as helpful as these organizations are, they cannot overcome inadequate funding, vague policy directives, or tedious acquisition procedures.

LAND EXCHANGE

In the absence of acquisition funds, land exchange offers the Forest Service its only other opportunity to consolidate national forest lands into efficient, manageable units.

On the Hoosier National Forest in Indiana, for example, the Forest Service exchanged 785 acres of federal land detached from

---

There are blocks of private land scattered within the Green Mountain National Forest that should be publicly owned. We have urged the Forest Service to exchange some of the land they own along the roads for those isolated inholdings. But exchanges are complicated, and the Forest Service feels it has to get a very good deal in the trade.

RICHARD CLARK, Selectman
Ripton, Vermont

the forest proper for a 731-acre inholding owned by two timber product companies in order to consolidate the national forest holdings.[33] Similar efforts have been underway on the Kisatchie National Forest in Louisiana to eliminate scattered fragments of public land and assemble a large nucleus of national forest land. Exchanges can serve other objectives as well. On the Ocala National Forest, the Forest Service discovered that a developer had purchased 172 acres nearly surrounded by national forest land, and planned to build a 525-lot subdivision. According to District Ranger Walter Guerrero, the subdivision would have been a visual intrusion on a large natural area, would have required access across national forest land, and would have imposed heavy traffic on forest roads not built to accommodate it. When the Forest Service's attempts to buy the 172 acres failed because its appraisal of value did not meet the owner's asking price, it proposed an exchange of 84 acres of federal land for the 172-acre inholding. "I consider it a good piece of business," said Guerrero. "The land the developer got was in three or four parcels near a developing community. It's good business to let small parcels like that go out of Forest Service ownership. And we got 172 acres which should have been part of the national forest."[34]

However, care must be taken to ensure that the Forest Service does not inadvertently facilitate a developer's assemblage of a large block of land that will result in an even larger, more damaging development on the forest periphery. The transfer of national forest land to a private landowner through exchange must be considered in light of potential future uses of the land and the effectiveness of local land-use controls.

Exchanges are procedurally complicated and time-consuming. Sometimes the Forest Service must arrange a three-way exchange, first acquiring land desired by the owner of the land sought by the Forest Service, and then trading it in turn to get the land it wants for the national forest. Further, the Forest Service's options are limited, since the law prohibits interstate land exchanges because they can affect an area's income from national forest revenue-sharing or, conversely, result in tax losses by removing land from the tax rolls. It is true that *large* interstate exchanges could create economic disruption in both the localities losing national forest land and the locality gaining federal acreage. But intrastate exchanges create the same problems for county revenues, and they are not restricted by law. **Congress should remove the present pro-**

hibition against interstate land exchanges by the Forest Service in legislation that would establish limits on the maximum acreage which could be removed within a given period of time from a single forest within a state. There should be no objection, for example, to the exchange of scattered parcels in the Ocala for private land in the Wayne National Forest in Ohio, so long as the forest whose land is used as trading stock is not significantly diminished.

RECEIPT SHARING

The fact that the Federal Government does not pay local property taxes on its national forest lands is the major reason for the characteristic resistance of county and local officials to Forest Service efforts to acquire land. This situation has also tended to sour virtually all relationships between counties and the Forest Service, in-

Ripton is a town of about 32,000 acres, and 23,000 are in the Green Mountain National Forest. This leaves a very narrow tax base for the town to operate on. If we are going to keep these nice open forest spaces for people, who's going to pay for them? How is the Federal Government going to compensate the towns for the tax loss?

I am from Chicago, and before moving up here I lived in New York. I know about the recreational and open-space problems of big cities. But now I'm on the other side of the desk. I don't have a park problem; I have a cash-flow problem.

Let's be candid about it. Why should the people up here be concerned about the people in Boston? They don't have to drive to get to a forest; all they have to do is step out the back door. Naturally, they're going to be concerned about the things that affect them most directly. As far as the Green Mountain is concerned, our residents care most about taxes and their own ability to remain in the area without being taxed off their land.

My view is this: I don't care what the Forest Service wants in the way of more land for the national forest. They are not going to get any more in Ripton. I have to ask, "How much of the town can we afford to have taken out of taxation?" It's great to keep this land open and available to the public, but is the public willing to pay for this land? Someone is going to have to.

RICHARD CLARK, Selectman
Ripton, Vermont

hibiting possibilities for cooperative action. "Something just has to be done about payments to counties," is a point of agreement among virtually all forest interests.

Finally, in 1976, the Congress addressed the problem with passage of P.L. 94-565. While inequities in returns to counties may persist and further revisions may be necessary as experience builds, the formula established in this legislation will pump thousands of additional dollars into the coffers of those timber-poor eastern national forest counties which have long looked upon these federal lands as a financial burden.

Counties have long received funds from the national forests. Under a formula established for national forests in 1908—even before passage of the Weeks Law—25 percent of forest revenues was returned to the states for distribution to the counties, with its use limited to support of schools and roads.[35] Each county received a share of revenues raised from the forest (or forests) within its boundaries—from timber sales receipts, recreation user fees, mineral leases,[36] and so on—based upon the amount of national forest acreage within the county. For a few, especially in the Northwest, this revenue was a bonanza; for others, including most Appalachian and Lake States counties, it was a pittance.

While some mistakenly labeled this process "payment in lieu of taxes," and it is officially termed "revenue sharing," it is more accurately referred to as receipt sharing, since the payment is related to forest receipts and has no correlation to the amount the land would bring into the county treasury if it were privately owned.

The problems inherent in the 1908 formula for payment to counties were recognized by the Public Land Law Review Commission (PLLRC) in its 1970 report, *One Third of the Nation's Land*. Commenting upon a study of revenue sharing and payment in lieu of taxes conducted for the PLLRC, it concluded: "The study made for this commission confirms the contention of state and county government officials that shared revenues amount to less than the revenues they would collect if the lands were in private ownership and subject to taxation . . . Although they were originally designed to offset the tax immunity of federal lands, the existing revenue-sharing programs do not meet a standard of equity and fair treatment to state and local governments or to the federal taxpayers."[37]

As the major revenue-producing forest product, timber has largely determined the amount each county receives.[38] Payments from national forest revenues to counties have varied widely nationally, ranging from $6.68 per acre paid by the Siuslaw National Forest in Oregon to the one-tenth-of-one cent per-acre payment of the Teton National Forest in Wyoming. While no eastern forests matched the Siuslaw or its sister Douglas fir forests in payments to counties, the Homochitto in Mississippi paid $3.43 in fiscal 1975; the Clark (now part of the Mark Twain) in Missouri, $2.07, mostly from receipts from mining of federally owned lead; and the Kisatchie in Louisiana, $2.01—substantial amounts when one considers that the median payment for all the national forests in the 48 contiguous states was only 16 cents per acre.[39]

There were three major difficulties with the former system. The perceived inadequacy of national forest payments in those counties where timber and other resource values are low caused local resistance to national forest acquisition. Further, there was little incentive for local governments to control land uses on the forest periphery, since local officials felt they had to maximize the revenue return from remaining private land.

Second, payments fluctuated according to a variety of factors (including the national economy, timber demand, and national forest land-use policy) beyond the control of local officials, and often the Forest Service. This frustration made long-range fiscal planning difficult, if not impossible. In the Allegheny National Forest, for instance, Warren County received $54,290 in 1971, $86,136 in 1972, $62,670 in 1973, $93,588 in 1974, and $92,246 in 1975.

Finally, because payments are linked to national forest revenue-producing activities, counties tended to encourage intensive use of the forest resources for activities that produced cash receipts. For example, a county in the Superior National Forest is supporting timber harvesting in the portal zone of the Boundary Waters Canoe Area; while another is supporting a federal lease for a copper-nickel open-pit mine a quarter of a mile from the BWCA.[40] Local governments often oppose wilderness designation because it would remove an area from timber production and prohibit tourist-attracting development.

Disagreement between localities and the Forest Service continues over the extent to which counties "lose" revenue as a result of federal forest land ownership. Usually forgotten is the fact that, in the early years of eastern national forest establishment, counties were begging the Federal Government to acquire land that was then a financial liability. Resource-rich counties undoubtedly receive more from receipt sharing than they would in property taxes if the land were in private ownership. Warren County, Pennsylvania, in the Allegheny National Forest, now receives more from national forest receipt sharing than it would from county taxes, according to County Commissioner David K. Rice.[41] In contrast, Itasca County, Minnesota, which includes 300,000 acres of Chippewa National Forest, estimates that at present assessed values and tax rates for privately owned forest land, the county would receive $822,000 if the national forest were in private ownership.[42] In 1975, Itasca County received only $41,180 in national forest receipts.

Local officials' claims of revenue loss because of public ownership are countered by Forest Service contentions that national forests draw tourist dollars into an area, localities do not have to service the public land, and the Federal Government relieves counties of some road-building expense and the cost of fire protection.

However, many Forest Service field-level personnel have long favored more equitable payments to counties. A typical comment: "It would make our job a lot easier."

Under the 1976 legislation, introduced by Representative Frank Evans of Colorado, virtually all counties will benefit, but some more than others. Counties can choose to receive either a flat sum per acre (up to 75 cents) in place of receipt sharing, or 10 cents per acre in addition to their receipt-sharing entitlement. In both cases, maximum payments will be linked to population on a sliding scale, from a maximum of $50 per capita for counties with a population of 5,000 or less to $20 per capita for counties of 50,000 or more people. Counties containing national forest wilderness areas that were designated after 1970 are to receive 1 percent of the fair market value of that area at the time the land was included in the wilderness system. Requirements that payments to counties be used only for roads and schools have been eliminated by the new legislation; the funds can be used for any governmental purpose. The 1976 legislation only authorizes the in-lieu payments. Congress must still appropriate the funds, estimated by the Department of Interior to amount to $118,000,000 annually for all categories of affected federal lands.

Since in fiscal 1975 only 12 of the 50 eastern national forests received receipt-sharing funds amounting to more than 75 cents per acre, most eastern forest counties will realize significant increases in their revenues. Those at the lower end of the scale, of course, will find their federal payments hiked substantially. For example, Craig County, Virginia, which received just over $5,000 in fiscal 1975, should receive about $80,000 under the new system.

This legislation should accomplish two objectives of reform: It will ensure additional money to poor counties and remove most of the uncertainty involved in county budget calculations, since counties will be guaranteed a per-acre payment irrespective of forest resource receipts.

While P.L. 94-565 goes a long way toward dealing with the financial problems of a number of very poor counties, it contains shortcomings that require future resolution. In particular, it fails to factor into the new payments equation the two benefits that public land managers most commonly cite as the advantages to counties from public land within their boundaries. They claim that national forests and other public lands tend to attract investment (tourist dollars and other related development), thereby boosting local economies. They also contend that the local governments' costs are reduced by the presence of federal lands which do not require costly public services. Further research will be necessary to find a way to quantify these hard-to-compute benefits so that they may be factored into the calculation of the federal in-lieu payment to each county.

The payment-in-lieu legislation may deter further acquisition, since Congress undoubtedly will consider not only the purchase price of the land, but also the continuing draw on the federal treasury for in-lieu payments.

Finally, P.L. 94-565 does not adequately address the issue of equity, which historically has been a major problem in the formula for receipt sharing. The new law will not really result in a more equitable redistribution of wealth between the resource-rich and resource-poor counties, although the spread between them will be lessened somewhat. But resource-rich counties will retain their advantage, and even realize a revenue increase, because of the 10-cent feature. Increases in timber harvesting or timber prices would mean continued disproportionate increases to those counties with timber wealth. As Forest Service Chief McGuire pointed out in testimony before the House Interior Committee: ". . . freezing the payment at 75 cents might tend to make a more equitable system over the first years, but over time there would be a drift toward continuing inequities because those not choosing 75 cents would continue to get greater receipts as time went on."[43]

Equity is both a philosophical and political issue. In principle, is it right for some counties to realize a bonanza from nationally owned resources, while other counties—less fortunate in climate, soil, and topography—find federal ownership a burden? Or, in other terms, to what extent should the Federal Government compensate those counties in which it has preempted the land and associated resources for national purposes?

Politically, any true redistribution of income would face bitter resistance from the "have" counties. Resource-rich counties cannot be expected to surrender their advantage gracefully in order to benefit counties in other regions of the nation. Given the present federal returns to southern pine counties, it is not surprising that a Florida county official acknowledged in an interview "the problem of equitability," but concluded that "the forests of one state should not subsidize the forests of another state."

The complex philosophical and political questions of fairness implicit in the payment-to-counties problem can never be definitively resolved. But experience with the new legislation and careful analysis of data should help correct continuing problems. Although it is an important and encouraging step, the passage of P.L. 94-565 should not be considered the final solution to the payments problem, but rather an interim adjustment, subject to future refinement.

REFERENCES

Chapter VI

1. Interview with John A. Sandor, Deputy Regional Forester, U.S. Forest Service, June 1975.
2. Data are from U.S. Forest Service, *National Forest System: Areas as of June 30, 1975* (Washington: GPO, 1975).
3. Interview with Donald Girten, Supervisor, Wayne-Hoosier National Forests, July 1976.
4. U.S. Forest Service, *Forest Plan, Allegheny National Forest* (n.p.: 1975), p. 31.
5. There is a measure of deception in some forest percentages, because forest-purchase boundaries often were drawn in the past to encompass large areas that include established communities and fertile agricultural valleys which never will be acquired.
6. U.S. Forest Service, "Mineral Considerations in Weeks Law Purchases and Exchanges," a report to the National Forest Reservation Commission (1972), p. 3.

7. William E. Shands, *The Subdivision of Virginia's Mountains* (Washington: The Conservation Foundation, 1974), p. 28.
8. U.S. Forest Service, *Plan for Managing the Ocala National Forest* (Atlanta: U.S.F.S., 1972).
9. Interview with Thomas A. Fulk, Forest Supervisor, Nicolet National Forest, February 1975.
10. Henry S. Graves, "Report of the Forester" (Washington: GPO, 1972).
11. Ibid.
12. U.S. Congress, *A National Plan for American Forestry* (Washington: GPO, 1934), p. 67.
13. Public Land Law Review Commission, *One Third of the Nation's Land* (Washington: GPO, 1970), p. 269.
14. Ibid.
15. Interview with David K. Rice, Chairman, Board of Commissioners, Warren County, Pennsylvania, July 1975.
16. Interview with John Humpke, Vice President, The Nature Conservancy, August 1975.
17. Board of Directors, Top of Alabama Regional Council of Governments, Resolution 72-23, October 31, 1972.
18. U.S. Forest Service, "Coal Evaluation Study, Monongahela National Forest," 1971.
19. Representative Morris K. Udall in an address before the National Wildlife Federation's 23rd Annual Conservation Congress, December 8, 1976, Washington, D.C.
20. U.S. Congress, House, "Surface Mining Control and Reclamation Act of 1975, Conference Report," no. 94-189, May 2, 1975, p. 85.
21. Uniformity between the federal and state criteria for unsuitable lands is important to buttress its legality, since unsuitable designation would not represent an arbitrary federal decision.
22. Letter from Dennis P. Grassi, Forester, Lands Staff, Sawtooth National Recreation Area, December 23, 1976.
23. Letters from Roy C. Gandy and Carl N. Wilson, U.S. Forest Service, January 1976.
24. U.S. Forest Service, "Open Space for People," in "Situation Report, 1973."
25. Letter from Carl W. Wilson, Director, Lands and Watershed Management, Eastern Region, U.S. Forest Service, January 1976.
26. Letter from Roy C. Gandy, Lands Specialist, Southern Region, U.S. Forest Service, January 1976.
27. Figures calculated from annual reports of the National Forest Reservation Commission for those years.
28. U.S. Forest Service, *Plan for Managing the National Forests in Missouri* (n.p.: U.S.F.S., 1975), p. 34.
29. National Forest Reservation Commission, *Annual Report, Fiscal Year 1975* (Washington: GPO, 1976).
30. The Conservation Foundation, *National Parks for the Future* (Washington: The Conservation Foundation, 1973), p. 13.
31. Figures are from the Bureau of Outdoor Recreation. The Park Service figure includes a special $65,500,000 appropriation for Redwood National Park.

32. Arthur A. Davis, "Alternative Futures for the Eastern National Forests: Problems and Prospects," prepared for The Conservation Foundation, December 1975.
33. Interview with Reginald W. Shepherd, Director of Appraisal, Purchase, Exchange & Donation Group, Eastern Region, U.S. Forest Service, November 1976.
34. Interview with Walter Guerrero, District Ranger, Lake George Ranger District, Ocala National Forest, April 1976.
35. Act of May 23, 1908. See EBS Management Consultants, Inc., *Revenue Sharing and Payments in Lieu of Taxes on the Public Lands*, report prepared for the Public Land Law Review Commission (Washington: GPO, 1970), vol. I, p. 18.
36. The 25 percent return from mineral leases pertains to acquired federal land. Public domain lands return 37½ percent of mineral lease revenues.
37. Public Land Law Review Commission, *One Third of the Nation's Land*, p. 236.
38. However, some eastern forest counties do realize a significant amount from leasing of federally owned minerals.
39. Data are from the U.S. Forest Service.
40. Cary B. Hinton, "Forest Service Payments to Counties: Minnesota," prepared for The Conservation Foundation, 1975. (Hereafter cited as "Forest Service Payments.")
41. Interview with David K. Rice, July 1975.
42. Cary Hinton, "Forest Service Payments."
43. Testimony of John R. McGuire in Hearing on Payment in Lieu of Taxes before the House Subcommittee on Energy and the Environment, 94th Congress, 1st Session, p. 320.

CHAPTER VII

the forests of the future

One of the original objectives in the establishment of the eastern national forests was to restore superior forests which could be emulated by private landowners. While the national forests still can provide useful lessons in land stewardship and environmentally sound management, their multiple-use mandate and the abundance of privately owned forest land in the East dictate that the national forests be managed much differently from surrounding private land. The national forests of the East must remain superior forests, but superior for new reasons and in new ways. To help achieve this, The Conservation Foundation has proposed two overarching principles to guide forest management as the twenty-first century approaches:

- On the eastern national forests, priority should be given to providing public benefits that cannot be supplied by private land, either because resources are unavailable or an economic incentive is absent.

- In managing its eastern forests for the long-term benefit of society, the Forest Service should give first priority to restoring them to the maximum attainable level of resource quality, emphasizing their potential as natural environments distinct from the man-made environments otherwise dominant in the East. The forest and its products should be used only to the extent that this continuing process of restoration is not interrupted.

The individual recommendations advanced in Chapters III to VI are designed to implement these long-term goals: new statutory direction for eastern forest management, improvements in forest and resource planning, the meshing of management of public and private land in ways that benefit both the forests and the rural regions in which they are located, and the encouragement of timber production and recreation on private forest land. Many of these recommendations deal with the interrelationship of the public forests and the private land in and around them.

What might the national forests of the East be like in the next century—perhaps in the year 2025—if The Conservation Foundation's recommendations were implemented? It seems useful to test some of the implications of these recommendations on three eastern forests—the White Mountain in New Hampshire, the Ocala in Florida, and the Allegheny in Pennsylvania.

These forests were selected because together they present a broad range of geographic, ecological, and resource needs requiring different management approaches: the protection of spectacularly scenic mountains and the accommodation of local growth in towns near the notches of the White Mountain; the accommodation of recreationists and timber production and gas and oil extraction on the Allegheny; and, on the Ocala, the balancing of heavy forest visitation with the maintenance of the environmental quality of a populous state with a multitude of resource problems. Most situations and opportunities facing the eastern forests are evident on at least one of these forests. The hypothetical application of The Conservation Foundation's recommendations to these three forests should provide a sense of how the recommended policies and programs might affect other eastern forests.

The three scenarios assume that the eastern national forests have been afforded special policy attention. First by administrative directive, then by congressional statute, the significance of these islands of public land has been recognized. This factor is a prime determinant of management direction. The statute, relatively short and to the point, established as policy the principle that the eastern forests are to provide forest products and recreational opportunities distinctive from those which are, or could be, supplied by the private sector. This principle has been incorporated in Forest Service guides for planning regions in the east and in individual forest plans.

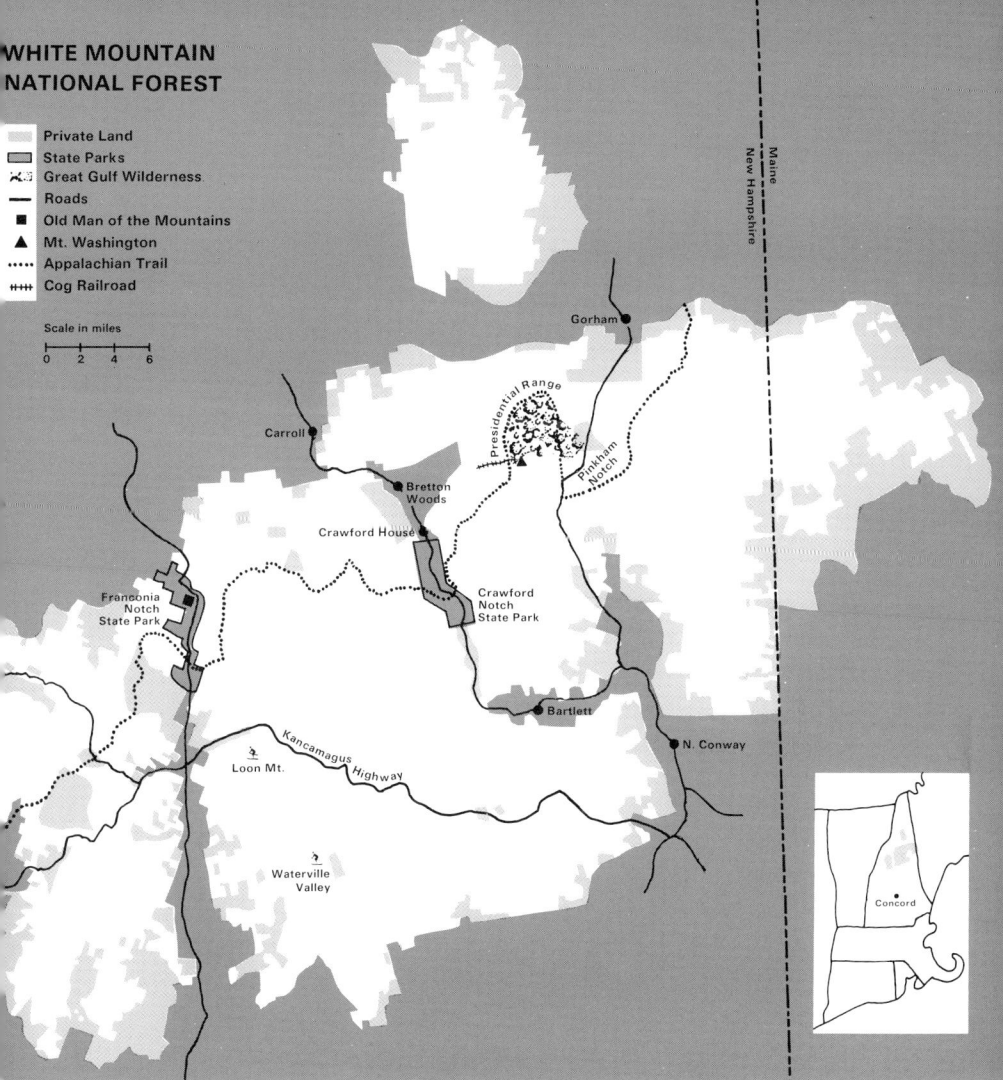

THE WHITE MOUNTAIN NATIONAL FOREST*

The 1974 Forest Plan for the White Mountain National Forest describes it as "one of the 'gems' of New England. Rugged mountains, deep forests, and associated rushing streams and rivers make this a recreation paradise for the urbanized East Coast."[1] The White Mountain National Forest consists of 683,637 acres in New Hampshire and 45,949 acres in Maine. In the White Mountain,

---

* Perry R. Hagenstein, executive director of the New England Natural Resources Center in Boston, Massachusetts, contributed substantially to this analysis of the White Mountain National Forest.

public ownership is more cohesive than in most eastern forests, with 80 percent of the land within the forest boundary federally owned. The most spectacular features of the White Mountain National Forest are the high peaks, particularly those of the Presidential and Franconia Ranges, and the three major north-south notches through the White Mountains—Pinkham, Crawford, and Franconia Notches. A large part of the forest was heavily logged by railroad in the first decades of the twentieth century; while traces of the old railroad grades are still discernible on the mountain slopes, much of the impact of the logging and fires that followed has been erased by lush, verdant vegetation. Today, timber harvesting is a modest activity on the White Mountain, although regionally significant. It occurs away from the high peaks and scenic areas of the forest.

The White Mountains have long been a focus for outdoor recreation in New England, especially in the summer. The White Mountain Forest Plan declares that the forest "is being managed under the multiple-use concept with emphasis upon mountain-oriented forms of recreation. At the present time, recreation has the highest relative value and is basic to the management programs for the forest."[2] A few of the rambling resort hotels, once served by railroads from New York and Boston, still remain. The forest is webbed with hiking trails, some a hundred years old. The Cog Railroad—"the old puffing devil"—was built in 1869 to carry tourists to the top of 6,288-foot Mount Washington, the highest peak in the northeast. Each summer, its engines still spew out black coal smoke on its trips to the summit. On the other side of Mount Washington is the Carriage Road, a privately operated toll road to the summit. In recent years, winter activities have gained increasing importance, with four major ski areas on the national forest and several others in the immediate vicinity. Winter hiking, climbing, and snowmobiling also attract large numbers of visitors to the White Mountains at a time of year when the forest used to be silent.

The long history of hiking in the White Mountains is built around a high-country hut system modeled after European alpine huts. Established and operated by the Appalachian Mountain Club, nine huts providing food and beds are strung along the most spectacular parts of the high country. The relationship of the Appalachian Mountain Club to the national forest is unique. Often suggested as models for other national forests, the AMC huts also have their detractors, because they tend to concentrate use. The last addition to the hut system was in 1966, although the AMC's 10.5-acre Pinkham Notch Camp (located on national forest land and operated under special-use permit) was expanded considerably in the early 1970's.

* * * * * * * *

In 2025, even a casual visitor to the White Mountain National Forest, approaching from the South through one of the three major notches, would notice something distinctive about the landscape. The national forest is immediately obvious, of course, since its superb timber and scenery distinguish it from the surrounding land, which is also largely forest. But more important, one also senses an integrity to the entire landscape, the kind of integrity that results from carefully planned decisions. In the small urban areas at either end of the notches—Conway and Plymouth in the south and Gorham and Littleton at the north—man and his developments are the dominant features. But 40 years of cooperation among the Forest Service, the State of New Hampshire, and some 20 towns in the area have managed to keep development in harmony with the most important feature of the region—the natural character of the White Mountain National Forest.

The groundwork for cooperative planning was laid in 1976, when the Forest Service acquired 475 acres of the old Crawford House property at the northern end of Crawford Notch, leaving in private hands 25 acres that included the sprawling Old Crawford House hotel itself. To ensure the integrity of the notch, the Forest Service knew it would have to work directly with local governments. A few years later, it took the initiative in helping the town of Carroll find ways to protect the area from development incompatible with the notch's natural and historic character, and in the process provided a model later emulated by other towns in and around the forest.

The Forest Service's cooperative action was stimulated by realization that land acquisition could no longer be used as a primary defense against intrusive development in the White Mountain National Forest. Other eastern national forests, which had much less cohesive federal landownership than the White Mountain, received priority for available acquisition funds. Maine and New Hampshire restrictions on Forest Service acquisition made future purchase prospects uncertain at best.

Concurrent with this was the Forest Service's adoption of a policy establishing the restoration and maintenance of natural environments as its primary management objective. It was evident to Forest Service planners that their efforts toward this objective would be undermined by the uncoordinated development of the Crawford House property and the much larger neighboring Bretton Woods property, both surrounded by national forest. Since acquisition could not be used to stop development, action had to be taken to guide development.

The selectmen in the town of Carroll were skeptical at first. With 40 percent of its land area already in the national forest, the town felt dominated by the Forest Service. However, federal legislation which provided for regular payments in lieu of taxes on the tax-exempt national forest lands had removed many of the political obstacles to town cooperation with the Forest Service. Forest Service planners, including specialists in community development assigned from the Milwaukee regional headquarters, offered technical help to the town. Albeit slowly, the townspeople came to view the Forest Service as a source of technical expertise and, indeed, innovative ideas rather than a burden to be borne by the town's taxpayers.

The solution to the problem posed by the Bretton Woods property was worked out jointly by the Forest Service, New Hampshire, Carroll, and two private organizations—the Society for the Protection of New Hampshire Forests and the Appalachian Mountain Club, both of which had a long-standing interest in the area. It amounted to development of a portion of the area in a way that minimized road and other costs to the town and required that new buildings be integrated into the landscape. Special attention was given to wastewater management, provision of national forest trail linkages to serve visitors to the town, and transportation access to the portion of the Bretton Woods property slated for development. Of particular help to the town was the ability of the Forest Service, the state, and private organizations to draw on experiences elsewhere. Soon, other towns began to clamor for similar assistance from the Forest Service. The Forest Service's new plan for the White Mountain includes a section on future opportunities for cooperative action with Carroll and other towns in the region. The Forest Service now facilitates the provision of technical assistance and planning and management grants by the State and Private Forestry branch, other federal agencies, and the state.

In the late 1970's, information assembled for the Renewable Re-

sources Assessment made it clear that the increasing timber needs of the nation required a national timber program emphasizing the underutilized capacities of small and medium-sized private ownerships. As a result, the Congress enacted legislation for a program designed to increase productivity on private forest lands. The complex program combined tax and financial incentives and provision of technical expertise through the Forest Service's State and Private Forestry branch and the New Hampshire Division of Forests and Lands on a scale not previously attempted. As a result of this comprehensive program, most privately owned forests throughout the White Mountain region now have the appearance of being managed efficiently. The forest has replaced the hill pasture as the typical working landscape of New England.

The regional landscape has changed visually in other ways, too. For instance, where the fringes of the national forest used to shade into the surrounding private forest lands, there is now a clear physical boundary. Although logging still takes place on the national forest, the emphasis on the longer rotations needed to achieve a more natural appearance contrasts with the intense management of surrounding private lands and makes the boundary between the two as distinctive as that between a pasture and a field of potatoes.

To achieve the goal of restoring the natural environment, which the Forest Service and its citizen advisory committees agreed should mean a "relatively" natural appearance, the annual timber harvest had to be reduced. Though the White Mountain National Forest had not been a major nationwide supplier of timber, it was important to local timber companies. To minimize the impact on the local economy, a plan was developed to reduce timber production in phases. This gradual reduction was designed to allow time for private landowners to institute intensive management of their lands to offset the loss of production on the national forest. The Forest Service stressed that the reduction in its timber production would be temporary; once the forest had matured, timber harvesting—of large, valuable trees—would resume at a level near, if not above, the 1976 volume.

Considerable argument persists over the economic and social impacts on the region's timber industry of the natural environment policy. Not enough time has elapsed to determine whether the emphasis on longer timber rotations on national forest lands will attract new mills to process the large, high-value timber and add more to the economy than those that rely on smaller timber from private lands. However, it is already evident that the strengthening of forestry programs on private lands has enabled the local timber industry to operate at about the same level as in 1976.

Those who drive into the national forest itself still find many signs of man's presence. Even with the reduction in the annual timber harvest, logging roads are as frequent on much of the national forest as on the intensively managed private lands. A network of these roads is still required to provide access to the timber. But on the public forests, the roads are less heavily used and more care has been taken—through design and siting—to fit them into the natural landscape. Even with this care and greater emphasis on the utilization of logging residues, it is obvious that these roads are used at times for logging. Occasional stumps and tops of felled trees are found scattered throughout the two-thirds of the national forest that is logged by selection cutting. The alert observer may note that the proportion of sugar maple, hemlock, and beech has increased somewhat at the expense of yellow birch and white pine, but the national forest is mainly distinguished by large trees, rare on private land.

On one-third of the national forest, timber cutting is prohibited. Some argue that timber in these areas is being wasted, but in fact only a fifth of this area could be logged and protected economically in any case. The high country along the Presidential and Franconia Ranges and associated ridges, as well as some 60,000 acres of lower lying areas, are in the "no-cut, natural" zone, both to protect scenic areas and because the timber cannot be harvested economically under sound environmental practices. Most of this land has been officially designated as wilderness.

Logging is not the only restricted use in the White Mountain National Forest. Damage to the natural environment caused by hikers and other recreationists has required restrictions on their use, too. The first recreation use to be totally banned from the national forest involved off-road motorized vehicles. But the natural environment goal was endangered by heavy hiking use as well, especially in the fragile, tundralike zone in the high country. Expansion of intensive recreation-use facilities, which would have attracted many more hikers, therefore was sharply limited. The Appalachian Mountain Club was permitted to add one new high-country hut in the northeastern sector of the national forest, but only because it provided a link between the older huts and a new chain built on private and state land north and east of the White

Mountains. The older huts still are used, but the Forest Service rejected all AMC requests for expansion. By building its new huts off the national forest, the AMC is attempting to deflect hikers away from the forest, or at least disperse use more widely.

By the late 1970's, increasing numbers of high-country hikers had worn the forest trails into knee-deep ditches. Erosion damaged the fragile, sparse vegetation. To control this, many trails, especially in the higher and fragile parts of the national forest, were hardened to withstand heavy use. On the other hand, a number of hiking trails in the forest's fragile areas were allowed to revert to vegetation, over the protests of those who used them most; these now are among the wildest parts of the national forest.

The policy of discouraging developed recreational facilities on the national forest has kept recreation use from climbing as rapidly as projected in the 1974 forest plan. No additional national forest land has been leased for skiing or other resort uses and expansion of existing facilities has also been limited. The ski resorts that existed prior to the new policy—Waterville Valley, Loon Mountain, and others—continue to be powerful magnets for the public, as are those on surrounding private land. Although these enclaves clearly are not compatible with the natural environment policy, their impact is limited to relatively small areas in the forest.

THE OCALA NATIONAL FOREST*

A coastal plain forest located about 30 miles north of Orlando, Florida, the Ocala National Forest stretches some 45 miles north from Deland, Florida, and covers about 20 miles at its broadest point. The Oklawaha River rises in Lake Apopka, south of the forest, and flows along the western and northern edge of the forest. Part of the Oklawaha has been channelized as a result of the aborted cross-Florida barge canal project. St. Johns River and its intervening Lake George form the eastern boundary of the Ocala. The most distinctive natural figures are the Big Scrub, a sandy, dry, 176,000-acre area that forms the central north-south backbone of the forest, and numerous swamps, lakes, and springs to the east and west of the higher Big Scrub.[3]

Forest Service statistics indicate that the Ocala is one of the na-

---

* William M. Partington, Executive Director of the Florida Environmental Information Center in Winter Park, Florida, contributed substantially to this analysis of the Ocala National Forest.

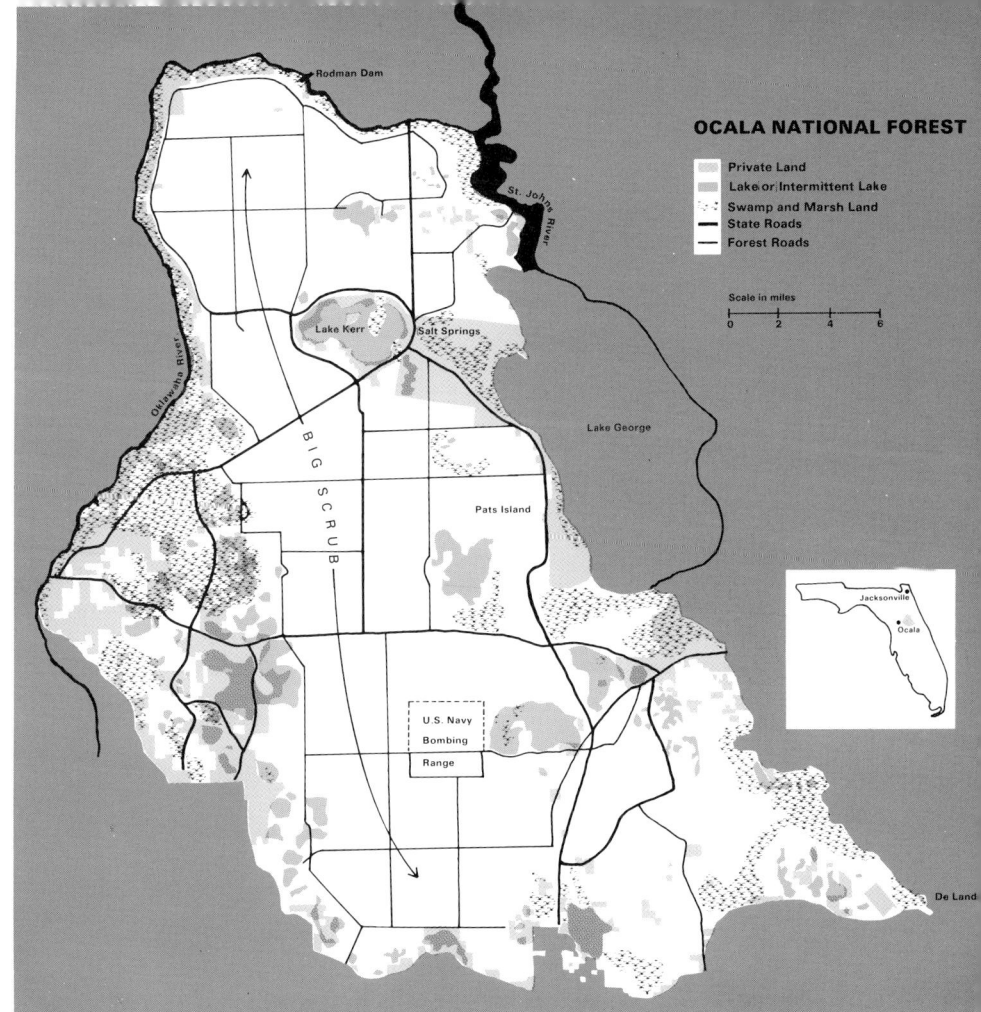

tion's most popular forests for recreation use, with more than 2,000,000 visitor days annually. Although it produces less timber than most coastal plains forests, accounting for only about 5 percent of the wood used by the local timber industry, it is important locally because the forest's sand pine and longleaf pine can be harvested during the wet season when other timber is inaccessible. The Ocala features a remarkable diversity of wildlife: the Big Scrub deer herd, squirrel, turkey, quail, and numerous species of animals and birds classified as rare or endangered, including the Florida panther, black bear, and bald eagle.

A nine-square-mile area in the center of the Ocala serves as a Navy bombing range and is closed to public use. Unlike most eastern national forests, the Ocala was established from public domain land, which means that public ownership is more consolidated than on most eastern forests. Eighty percent of the land within the

Salt Springs, a developed inholding in the Ocala National Forest.

forest is federally owned, and the Federal Government controls the mineral rights. However, extensive inholdings are scattered along the eastern and western edges of the forest. These are infiltrated by over 125 subdivisions, many of which are substandard and blight the forest scene. Additional private land surrounds many of the larger lakes, and there is a substantial inholding around Salt Spring, which averages 52,000,000 gallons of water a day.

The 1972 forest plan for the Ocala defines a principal objective: "to manage the Ocala National Forest in an apparently natural condition. Any practice that could hinder the attainment of this objective will be evaluated for impacts and alternatives." [4]

* * * * * * *

By the year 2025, Florida is focusing on ecosystem planning as a natural outgrowth of its implementation of federal and state laws of the 1970's. These include the federal Air- and Water-Pollution Control Acts, the state Land and Water Management Act, the Water Resources Act, and the Local Government Comprehensive Planning Act. Ecosystem planning was in large measure stimulated by the threat of water shortages in the populated areas of central and southern Florida and the contamination of groundwater supplies from saltwater intrusion, sewage, and chemical wastes resulting from development. In the region surrounding the Ocala, the ecosystem planning program that developed was facilitated by new Forest Service cooperative action programs.

Ecosystem planning first focused on land and resource capabilities and regional "carrying capacities." A resource capability study of the Oklawaha regional ecosystem, similar to earlier studies of the Green Swamp and South Florida, was carried out by energy systems analysts and planners at the University of Florida. The capability study was incorporated into the plans of regional planning councils and water-management districts. Since the Ocala National Forest is a major regional resource, the Forest Service was a principal cooperator in the planning process and the regional study was incorporated in the Ocala forest land-management plan.

Water shortages in such populous areas as Tampa Bay, the entire developed portion of the Atlantic Coast, and in Orlando had at first encouraged exotic schemes to provide water, such as desalinization or the transport of water across the state from springs on the Gulf Coast. But even the coastal springs became turbid (one of the first was Weeki Wachee in 1976) or high in chlorides, and the importance of high-quality groundwater from areas such as the Ocala National Forest and the Summit Pool section of the old barge canal route demanded protection of the Ocala region's water resources. A water-supply budget study that considered net recharge (rainfall less transpiration, evaporation, and runoff) showed that significant withdrawals could be made from the Ocala National Forest without depleting the resource. Consequently, the Ocala has become an important supplementary source of water for many coastal communities. To keep the regional water loss to a minimum, wastewater receives sophisticated treatment and is sprayed on new-growth scrub areas in parts of the forest, both stimulating tree growth and recharging groundwater. Strict standards regarding suspended solids, nitrates, chlorination, and coliform bacteria are enforced to prevent groundwater contamination. Water-supply wells in the forest are carefully monitored.

Because withdrawals are based on a carefully controlled water budget (central Florida, like other areas of the state, has been forced to limit growth consistent with water supply), the Ocala's numerous springs continue to produce high-quality water. The problems of the weed-choked Rodman Reservoir on the barge canal route have been eliminated. Those areas of the barge canal route adjacent to the Ocala National Forest were retained in public ownership and incorporated into the Ocala; restoration to near-natural hydric conditions is well underway.

A massive restoration program of the upper Oklawaha was begun in the late 1970's, with a draw-down of Lake Apopka. Other lakes downstream were progressively restored, as were channelized portions of the upper Oklawaha. The Oklawaha is once again a largely self-maintaining river, since the water levels fluctuate naturally, and there has been a considerable improvement in water quality. All of the Oklawaha adjoining the national forest, from Silver Springs Run to the St. Johns River, is classified as "scenic" under the federal Wild and Scenic Rivers program, and may soon be upgraded to "wild." The objective in improving and protecting the river was not only to restore scenic values; it was recognized that a naturally functioning river was vital to the health of the entire regional water regime.

To help restore the forest's natural environment, the Ocala's road network has been given considerable attention. Roads that formerly paralleled straight rectangular survey section lines have been reseeded, and a network of unpaved Forest Service roads has been established for official access. They are also used—under short-term permit—by hunters and other members of the public. These roads wind to avoid sensitive areas and discourage high-speed travel, but permit Forest Service access to most areas vulnerable to fire. The old straight highways that invited motorists to cross the Ocala for convenience, and at high speed, have been closed or modified to discourage such use. Now roads only provide access routes into—but not through—the forest.

Multiple use is still practiced on the Ocala National Forest, but in the context of the regional plan. Both forestry and grazing are used as ecosystem management tools. Limited grazing is permitted in a few fenced portions of the Big Scrub where there is a high growth rate because of fertilization with sewage nutrients. This is permitted as a natural recycling of the waste nutrients.

The Ocala's Big Scrub remains the primary source of commercial sand pine. However, some private landowners again are growing this species in well-drained areas where citrus is susceptible to freezing. Some clearcutting of sand pine continues, providing local pulp mills with a source of wood during the wet summer season when pine flatwoods are flooded. Clearcutting occurs only where studies have shown there is little threat to ground or surface waters, unusual wildlife species, or recreation. Some clearcutting helps much of the area's wildlife and regenerating areas are shown to visitors and students as examples of forest succession. Cutover scrub areas are also planned to serve as firebreaks.

Private landowners now find timber production more attractive. Those who grow trees for sawtimber on rotations longer than 50 years can obtain insured low-interest loans against the timber so they do not have to sell to meet temporary financial emergencies. Wastewater effluent is sprayed on private forest land to lengthen the growing season, permitting timber intended for pulp to be cut profitably after 20 years. Growers of pulp and sawtimber prefer only moderately improved strains. Genetically superior trees are selected more for their resistance to pitch canker or other diseases than for their growth rate, since the fertilizing effects of the effluent make growth extraordinarily rapid in any event. There is greater recognition that genetically diverse trees, within any species, afford greater protection against diseases or other plagues.

Because the economic values of water quality and wildlife have been quantified, private timber growers give more consideration to environmentally sound management than in the 1970's, when existing methods of computing costs and benefits favored the hard-cash returns from timber. The economists were able to show that

land managed in environmentally sound ways provided public benefits for which landowners should be compensated. Now the privately owned flatwoods north, west, and east of the national forest are used both to grow timber and to provide a broad range of environmental benefits.

An incentive is provided through an "environmental exemption" on the landowner's property tax. These exemptions vary with the use of the land, its favorable impacts on water and air quality, the land's ability to utilize wastewater effluent, and other factors. The actual tax rate also reflects the probable ability of the owner to have invested elsewhere for higher returns, as well as anticipated profits from timber sales. Such a complicated system would have been thought impossible to administer in the 1970's, but the prospect of a deteriorating environment made extraordinary measures imperative. The methodology was developed and tested initially to provide incentives for preserving prime farmland, which became the critical environment-resource issue in the late 1970's and early 1980's.

Lower property taxes have only partially mollified rural landowners, who found their paper profits drastically reduced by tough state and local land-use regulations. With intensive development foreclosed, new buyers of land outside town limits are coming to value it for its productive potential or its use as a homesite rather than for the possibility of future subdivision.

The acquisition program in the Ocala has focused on the Oklawaha River, and the Lake Kerr-Salt Springs areas. The Forest Service purchased Salt Springs, a 10,000-acre area once privately owned and operated as a tourist attraction. One of the major springs in the forest and central Florida, it has been replanted by the Forest Service with native vegetation and permitted to return to a more natural state.

Substandard subdivisions built on inholdings and private land adjacent to the forest, particularly in and adjacent to the fire-prone Big Scrub, have also been the subject of an aggressive Forest Service acquisition program. The Forest Service acquired two of the worst subdivisions, condemned by local officials as health hazards; removed most of buildings; and, capitalizing on the existing road network, operate them as overflow campgrounds during hunting season. Land exchanges have consolidated public ownership at the western edge of the forest and eliminated many of the private inholdings. County restrictions on signs, setbacks, and building codes have improved the appearance of those that remain. But there are still many inholdings, particularly since the Forest Service acquired land along the Oklawaha River, west of land that continued in private landownership. Environmental tax exemptions and scenic easements have encouraged most homeowners to replace lawns with native groundcover, usually by natural succession. Several owners willed their properties to the Forest Service for tax benefits; these are being restored to a natural condition.

The Navy bombing range has been phased out. Some larger buildings were retained for Forest Service use, but most were removed. A portion of the fenced range considered relatively clear of unexploded bombs became an experimental cattle range. However, the possibility of unexploded bombs keeps the area off-limits to forest visitors.

Carefully executed exploration for oil in the national forest indicated that only minimal reserves existed in the Ocala. These reserves, federally owned, were considered relatively insignificant in terms of national supply. In keeping with the national energy policy developed after the second oil embargo by the Organization of Petroleum Exporting Countries, the Ocala reserves are being held in case of an acute emergency.

Continued national uncertainties in energy supplies and higher fuel prices have curbed the growth of high-intensity recreational developments designed to attract a national clientele. Many tourists still come to Florida, of course, most by public transportation that now includes an extensive system of "auto-trains" modeled after the highly successful Washington-to-Florida trains initiated in the early 1970's. But nationally, the trend is toward recreation within one's own region, and this is true of Florida as well. Floridians are increasingly looking to opportunities for many different kinds of recreational opportunity within the state.

The Ocala continues to be one of the most popular national forests in the South. The Forest Service estimates that it accommodates about 4,000,000 visitor days a year, double the 1970 statistic. However, most visitors spend only the daytime in the forest. Numerous private campgrounds outside the forest provide comfortable, secluded campsites. Through an experimental program, the Forest Service recommends private campsites which meet stringent health, safety, and environmental standards. While this program was at first controversial, with some owners objecting to what they viewed as federal regulation of private activity off the forest, state and local health and environmental regulations have achieved much the same objective. More visitors now turn to private campgrounds because the Forest Service has increased the per-night rental fee for the few developed campgrounds on the national forest to a level about equal to that of the private campgrounds. Within the national forest, a new system of primitive "pack-in" campsites provides a camping experience for those willing to leave their cars and walk a mile or so into the woods.

On the national forest itself, recreation is dispersed. Although the Big Scrub is attractive mostly during the cooler months, an interpretive program that equates it to a "wet desert" has increased its year-round appeal. A network of trails permits visitors to see forest features such as the longleaf pine "islands" (distinctive areas in the slash-lobolly pine forest), sink holes, hardwood hammocks, wet prairies, ponds, streams, and springs. In low areas that are periodically innundated, such as parts of the Oklawaha Valley, wooden boardwalks have been constructed, some to mid-tree heights, permitting visitors to see the superlative wetlands communities in all seasons. Interpretive signs line the walks, and detailed guidebooks can be purchased.

Many prefer motorized transportation to tour the forest. Small, nonpolluting buses offer round trips into the forest—using roads not open to the public—at reasonable cost.

A mid-1970's study by the Florida Committee on Rare and Endangered Plants and Animals indicated that some 400 species native to Florida deserved special consideration. An outgrowth of this study was management of the forest to protect the habitats of many of the smaller mammals, the scrub jay and smaller birds, the scrub lizard, certain snakes, and several arthropods that occupy special niches within the forest.

As a result of decreased motorized activity within the forest and the reduction of intensive areas of human activity, the panther and black bear seem to be returning. At least they have not disappeared, as it appeared they might in the 1970's. Restocking of the Ocala with panther from captive stock was considered but rejected, since few semi-tame animals adjust to life in the wilds.

Bald eagles, swallow-tailed kites, ospreys, limpkins, egrets, and other large birds can be observed from boardwalks that cross wet areas. The red-cockaded woodpecker, seasonally quite numerous, is readily seen in the longleaf areas. Its nesting trees are marked for easy identification. The Forest Service promotes the fact that only

in this forest can one expect to see this particular mixture of wildlife, which includes several species found in only a handful of counties outside the Ocala region.

Hunting continues on the Ocala under rather strict controls. Through careful regulation of the quota system for hunting permits, first instituted in the early 1970's to control hunting pressure on the herd as well as for hunter safety, the Ocala's deer herd increased to the 10,000 animals projected in the Forest Service's 1972 forest plan. The Florida Game and Fresh Water Fish Commission adjusts the hunter-permit quota each season on the basis of estimates on the size and health of the herd. The seasonal hunting permit quota has fluctuated between 5,000 and 7,000.

There have been other changes as well. The historic Florida practice of hunting deer with dogs has been strictly regulated. In order to separate dog hunters from other hunters and forest visitors, the use of hunting dogs is restricted to specific areas of the forest on special days during the deer season. Pressure for the restrictions came particularly from hunters who prefer to hunt deer on foot without noise from vehicles and dogs. The elimination or closure of many national forest roads also served to inhibit this type of hunting, because hunters' vehicles could no longer easily follow the dogs.

Hunting on the Ocala is now pretty well limited to dedicated hunters willing to pay the increased fee—which is high even by inflated 2025 standards—and who are willing to park on the periphery of the hunting area and walk in to seek their quarry.

The higher fee for hunting on federal land and the limitation on national forest hunting permits has given a number of large landowners in central Florida a stronger incentive to lease hunting rights to gun clubs and other groups, as well as to the state, which annually publishes a directory of lands open to hunting. The state's Game and Freshwater Fish Commission advises private owners on managing and increasing their deer herds.

The natural environment policy has enhanced the natural features of the forest and attracted visitors interested in recreational opportunities that emphasize education and ecological understanding. Timber harvesting remains about as important a forest activity as it was in the 1970's, but forestry is regarded as an integral tool in the management of the forest's wildlife and hydrological system. The benefits of integrated ecosystem management extend far beyond the forest boundaries.

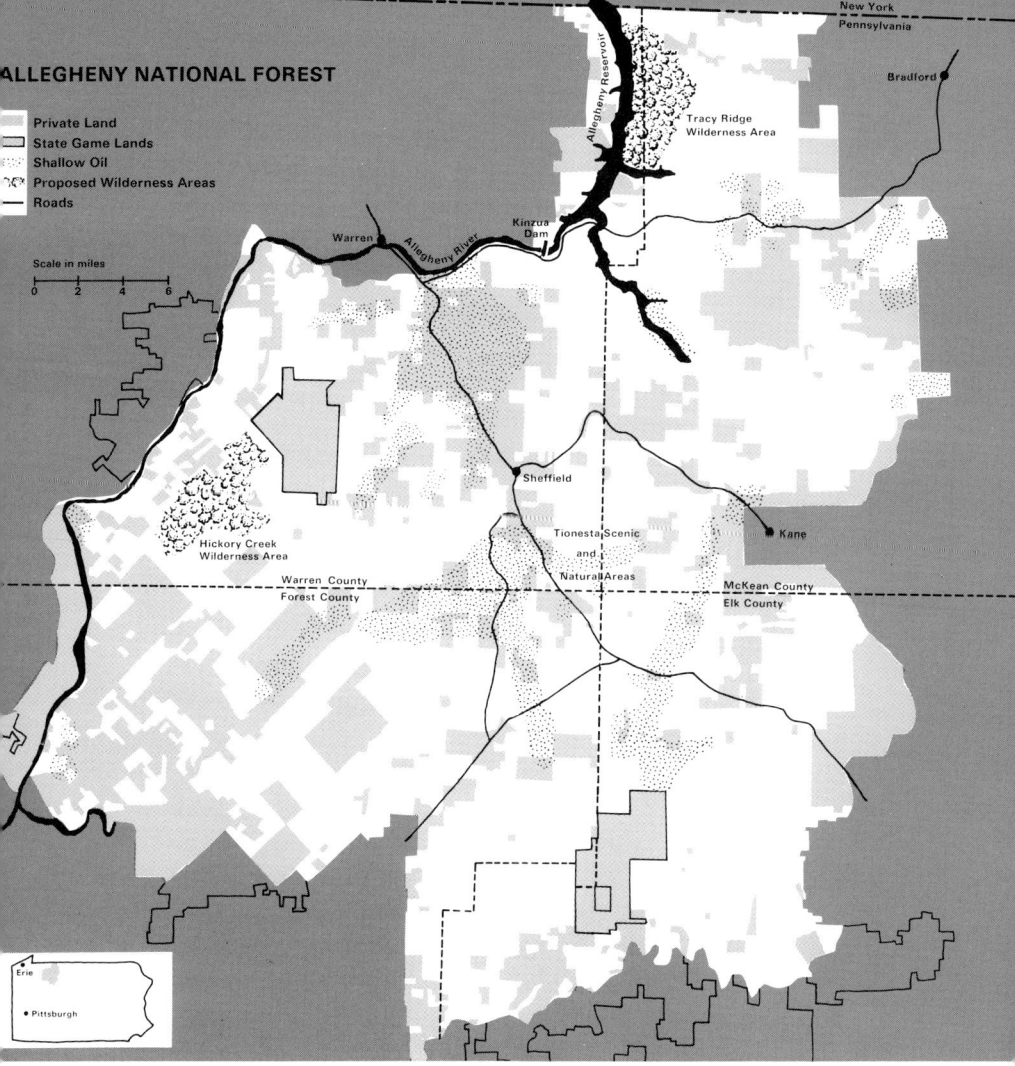

THE ALLEGHENY NATIONAL FOREST[*]

Conflicts inherent in the concept of multiple use are apparent on the 501,857-acre Allegheny National Forest, spread across the rolling Allegheny Plateau in Pennsylvania's northwestern corner. Large and growing numbers of users value the forest for the amenities it offers, both for recreation and as a natural environment. But equally significant are the many who rely on the forest as a source of timber essential for jobs and the economic stability of the area's towns and communities.

---

[*] Arthur A. Davis, Director of the Pennsylvania Land Policy Project of the Western Pennsylvania Conservancy in Pittsburgh, Pennsylvania, contributed substantially to this analysis of the Allegheny National Forest.

The Allegheny ranks among the most timber-rich of the Appalachian forests, with valuable stands of black cherry, ash, maple, and oak.[5] Its high-quality lumber and veneer material serve a national market.

But timber is not the only commodity. Beneath the forest surface lie rich pools of Pennsylvania crude oil, highly valued for its properties as a lubricant. The Forest Service estimates that there may be up to 20,000 barrels of oil per acre beneath the forest. But very little is owned by the Federal Government; on only 1 percent of the Allegheny has the Federal Government purchased subsurface mineral rights. The remaining mineral rights are outstanding, which means the owners are entitled to "enter undisturbed" upon forest land to exploit the minerals.

The Allegheny's location undoubtedly contributes to its popularity as a recreation area. It lies at the hub of the eastern seaboard and Great Lakes population centers, with 92,000,000 persons living within a day's drive of the forest. Within 50 miles of the forest reside more than half a million persons, projected to increase to three-quarters of a million by the turn of the century. The forest serves a variety of recreationists: boaters and swimmers on the Allegheny Reservoir; campers, hikers, and backpackers; and snowmobile and other off-road vehicle enthusiasts. In 1975, the forest recorded about 2,000,000 visitor days of recreation use. Use of the forests by off-road vehicles—snowmobiles, motorcycles, and four-wheel drive vehicles—is an increasing, and controversial, recreation use of the forest. While the Allegheny does not boast the spectacular scenic features of the White Mountain, three areas have been proposed for inclusion in the National Wilderness Preservation System.

The most abundant animal in the forest is the white-tailed deer, whose grazing has impaired timber regrowth in some areas. While one is most likely to see deer, there are, according to the Forest Service, 45 other species of mammals, 132 species of birds, and 71 species of fish found in the area. The more unusual wildlife includes the river otter, bobcat, and raven. The black bear also inhabits the forest, and the Allegheny accounted for 12 percent of those taken by hunters in Pennsylvania in 1973. Turkey is also a prized small-game species. Six species of rare or endangered fish are found in the Allegheny's waters.

The prevention of flooding in Pittsburgh and other downstream cities and towns was one of the major reasons for the establishment of the Allegheny National Forest. Today, the Forest Service calculates that just over half of the 800 billion gallons of water that fall on the forest annually flow into area streams.

In charting its 1975 direction, the Forest Service attempted to reconcile the many interests in the Allegheny:

> . . . the forest will remain a working forest, producing forest products that make a fair monetary return to national, state and local treasuries, in addition to supplying materials to meet many of man's basic needs. The forest will remain open to use, a place where mind, body, and soul can be refreshed through close association with nature.[6]

* * * * * * * *

In 2025, the Allegheny does not face the intense pressures and conflicts other national forests are experiencing. Although 100,-000,000 persons live within a 300-mile radius of the forest, that figure does not offer a meaningful gauge of pressure. People in Boston and Chicago do not tend to spend their leisure time in the Allegheny. The interstate system has not been expanded much since the Bicentennial days, travel costs are high, and many more park and recreation opportunities have been developed closer to home. Western Pennsylvania is still off the beaten track. Visitors come either because the forest is particularly convenient to Pittsburgh and Cleveland, or because it boasts a large area of unspoiled country, with high natural quality and landscape values. The forest staff attributes its major management problems to differences between these two principal clienteles. The visitor who comes because the forest is a close place to fish and trail bike has different expectations from the forest than someone who makes the journey to find refreshment in a natural setting of size and integrity.

People management replaced timber management as the national forest's biggest crop long ago. In response to the large number of recreationists, the Forest Service now dispenses solid information about what a forest is, how it grows, and how to enjoy it. And there is much more control over forest recreational activity. The Forest Service works closely with state and local officials to eliminate roadside camping and clutter. Trail bikes, snowmobiles, and other off-road vehicles must use designated trails.

Pennsylvania still sells more hunting licenses than any other state, and the forest has a strong demand for conventional hunting —bear, turkey and, of course, deer. Because of greater use of selection cutting in the harvesting of timber—and fewer clearcut areas —there are fewer deer in the forest than in the 1970's. But because the state has purchased additional game lands and instituted a new hunting program on leased private land, hunting opportunities in the region have remained stable. The forest charges less for hunter access than private landowners do, but the success ratio is lower on the forest.

Sportsmen think there is no substitute for killing a deer or catching a fish, but probably as many people click a shutter as pull a trigger. Other visitors come to the forest for birding and nature study. The inherent conflict between the two groups continues, but the Forest Service tries to minimize it by designating more natural areas on the one hand, and intensive game-management areas, such as recently cut timberlands, on the other.

Wilderness, too, has its fascination—and is still the center of much controversy. Values have changed, of course, but the notion of designating an area to go to seed or waste still outrages many, just as the knowledge that some small corner of the world will remain untouched and unspoiled comforts and nourishes others.

In any event, there is little to quarrel about. The two areas formally designated as eastern wilderness in the 1980's—the 7,900-acre Tracy Ridge Wilderness and 10,000-acre Hickory Creek Wilderness—see their quota of use, and satisfy the wilderness advocates as nothing else can. However, many others are just as pleased with the "low-intensity areas"—primitive, natural areas that differ from the designated wilderness areas in one respect only. At the conclusion of long rotations, the prime timber is selectively logged, the area redesignated as a general forest-management unit, and another area, set aside for the purpose years before, is selected to replace it. These rotating near-wildernesses are favored by some, who find that the remains of logging roads make access easier, and travel is not made treacherous by blow-downs and burn-areas thick with brush and greenbriars. The character of the old-growth timber and the degree of privacy and solitude meet the needs of most.

The fees charged for wilderness or near-wilderness use cover both the cost of management and lost production opportunity costs. The designated wildernesses are more expensive to use, of course, since all chance of timber income is foregone.

Those who seek something less primitive than can be found in the wilderness gravitate to one of the modest, well-designed rental accommodations which private landowners have found it profitable to build near the forest, on sites specified by local jurisdictions. Camping is still important, but the days of recreational vehicles are past. They became too expensive to power. Mini-buses operated by concessionaires take visitors to private campgrounds in the area, and to the few on the forest. Camping gear can be rented reasonably in neighboring communities. So can bicycles, and their popularity has stimulated requests that forest trails be "hardened" so bikes can use them. The Forest Service has resisted this, but bicycles are common on the better forest roads.

Pennsylvania was once the most forested of all the states; at least 97 percent of it was woodland when the first settlers arrived. Indiscriminate cutting and burning reduced this area to about 30 percent by the mid-19th century, but by 1975 over 60 percent of the Commonwealth had been returned to forest, and the acreage continued to increase, most of it in small, privately owned parcels. In that year, the state grew three times as much wood as it cut. Beginning in the late 1970's, state policy promoted intensive management of these small, private holdings. The federal and state governments encouraged the establishment of forest cooperatives, and supplied technical and financial assistance. Research developed methods of utilization and equipment tailored to small forest ownerships and to people who wanted to minimize aesthetic damage. Thus, although many owners did not want to harvest timber—and still do not—production of timber on private lands skyrocketed. The practice of timber management on private lands was aided by the heavy demand for firewood which paid the landowner for thinnings and other stand improvement.

The Allegheny now grows timber on longer rotations, and leaves entire areas unharvested. National forest management has focused on the black cherry, select ash, and one or two other hardwoods, depending on the fashion, used in furniture manufacture, paneling, and other specialty products. Management has almost returned to the forestry practiced by the Germans six centuries ago when they produced the *geldbaum*—"gold trees." Even though much of the forest is in older age classes—some has not been disturbed for nearly a century—people still find fascination in the virgin timber sites at Tionesta and Heart's Content and visit them in large numbers.

Developments over the past 50 years have stimulated considerable change in the attitudes and policies of state and local governments toward the national forest. The major changes stem from a new sense of the interdependencies of regions and institutions. This constitutes a strong argument for local, state, and federal cooperation.

The State of Pennsylvania used to give little attention to the Allegheny National Forest. To be sure, state agencies enforced environmental control laws to clean up oil and gas pollution on the forest, and police helped the Forest Service with traffic and law-enforcement problems. However, the state largely regarded the forest as a federal responsibility. While state officials and the Forest Service were cordial, there remained a certain concern for protecting one's own turf, a certain wariness about relationships.

At the local level, the difficulties were probably even more severe because of the uncertain income local governments received from the federal lands, and other problems arising from having the Federal Government the major owner, or the major influence, in a local jurisdiction.

Today, far fewer collisions and confrontations occur among various levels of governments. The sorting out of intergovernmental relations has made each level of government more secure in its own role. The stabilization of population and fair progress toward handling the major social and economic problems have helped. The emergence of a strong state land-use strategy also contributed to new initiatives in state-Forest Service cooperation. Some direct results include:

- The Allegheny River Boat Trail, a series of island and shoreline acquisitions from Warren, Pennsylvania, to Pittsburgh, was planned and put together by the state, the Forest Service, and the Western Pennsylvania Conservancy. The idea originated with the Conservancy, which helped acquire the land in advance of state purchase. The state bought the land, installed the basic facilities, and now conducts the annual program of environmental education and outdoor recreation on the river. Forest Service cooperation was the key to adding the headwaters area to the system.

- State and federal wildlands trails are harmonized into a regional system all the way from the Allegheny to the Susquehanna. The North Country Trail leads to the Susquehannok Trail with no interruptions, for example. And on the ground, no distinction is evident between federal and state segments of the trail. Design, facilities, even signs are the same. No longer do slippery logs span state waters while hewn-beam bridges cross federal streams.

Coordinated land management extends across most of the northern third of the state, from the Allegheny River to the Poconos. The state manages its northcentral highland holdings—7 state forests, 15 state parks, and 17 state game lands, totaling over 1,300,000 acres—according to a land-management agreement it entered into with the Forest Service in the 1980's. Timber sales, for example, are coordinated for the federal and state land to ensure a continuous supply to meet local needs.

The indirect consequences of this regional cooperation, while less dramatic, are probably more important. Land-use planning and development in the Allegheny region have been effective ever since the Congress agreed to a formula for paying local governments in lieu of taxes on the federal forest lands. Localities recover a comfortable minimum return from all federal lands, easing the pressure on land jurisdictions to permit high density and obtrusive development close to the forest.

Equally important, the state decision to make counties the focal points for land-use planning and regulation relieved the Forest Service of having to coordinate its plans with every village, borough, and township individually. The counties now have the funds and expertise to do a competent planning job, and most of the early difficulties and suspicions have been laid to rest.

The benefits are visible enough. Second homes and private recreation facilities come under state site and design standards to make sure they meet environmental requirements. The Forest Service participated in establishing the standards. All private oil and gas development is under strict state control, and the Allegheny's oil production is earmarked largely for lubrication uses and for production of the new synthetics.

In the 1980's the Forest Service began buying up oil and gas rights beneath selected areas of the forest and has consolidated its subsurface ownership along Kinzua Creek, which drains into the Allegheny Reservoir. The Forest Service now carefully controls extraction of the federally owned minerals to prevent soil erosion and pollution of the stream from the oil-drilling operations. When surface and subsurface ownership is divided, the state monitors as vigilantly as the Forest Service; by tacit consent, the state takes the lead in the enforcement of regulations.

* * * * * * *

Fanciful futurizing? Perhaps. But many aspects of these three future forest "visions" are based on a study, finding, or initiative being developed today in one of these forests. The Conservation Foundation advances these to illustrate the opportunities for imaginative management, sensitive to the special features of each forest and the needs of its region.

The Conservation Foundation is convinced that the nation faces in the near future some very difficult choices about resource use. An example already at hand: whether to divert water and land in the Plains States from agriculture to coal extraction. So far, the eastern national forests have not had to confront such a dilemma, but trends in increased recreation use, demand for timber, and energy needs make similar situations inevitable.

The eastern national forests of the future may not resemble the visions presented here. But they are not likely to resemble the future as depicted in forest plans either. The future is not a straightforward march to a long-established objective, but many waverings with frequent mid-course corrections. Those waverings can be simply random wanderings toward whatever attracts attention at the moment, or they can be constructive, goal-oriented adjustments to changing circumstances. Is it better to have the fate of the eastern national forests shaped by the vagaries of the present, or structured in accordance with a vision of the future? The latter course seems eminently preferable.

REFERENCES

Chapter VII

1. Data on the White Mountain National Forest are drawn from U.S. Forest Service, *Forest Plan, White Mountain National Forest* (Milwaukee: U.S. Forest Service, 1974).
2. Ibid., p. 9.
3. Data on the Ocala National Forest are drawn from U.S. Forest Service, *Plan for Managing the Ocala National Forest* (Atlanta: U.S.F.S., 1972).
4. Ibid., p. 13.
5. Data on the Allegheny National Forest are drawn from U.S. Forest Service, *Forest Plan, Allegheny National Forest* (n.p., 1975).
6. Ibid., Preamble.

APPENDIX A

STATISTICAL PROFILE OF THE EASTERN NATIONAL FORESTS[1]

STATE	Date Established	Gross Acreage Within Boundaries	Net Acreage	Net as % of Gross	Net NF as % of Total State Area	Administration
ALABAMA						
William B. Bankhead	1/15/18*	348,917	179,294	51%		National Forests in Alabama
Conecuh	7/17/36	171,177	83,957	49%		P.O. Box 40
Talladega	8/31/36	735,181	364,428	50%		Montgomery, Alabama 36101
Tuskegee	11/27/59	15,628	10,778	69%		
Total		1,270,903	638,457	50%	2.0%	
ARKANSAS						
Ouachita	12/18/07*	1,961,275	1,330,450	68%		Box 1270, Federal Building
(see also OKLAHOMA)						Hot Springs National Park
Ozark	3/6/08*	1,501,811	1,109,317	74%		Arkansas 71901
St. Francis	11/8/60	29,608	20,946	71%		P.O. Box 1008
Total		3,492,694	2,460,713	71%	7.4%	Russellville, Arkansas 72801
FLORIDA						
Apalachicola	5/13/36	632,589	557,729	88%		National Forests in Florida
Ocala	11/24/08*	430,378	367,204	85%		214 South Bronough Street, Box 1050
Osceola	7/10/31	161,814	157,230			Tallahassee, Florida 32302
Total		1,224,701	1,082,163	88%	3.1%	
GEORGIA						
Chattahoochee	7/9/36	1,640,915	741,279	45%		P.O. Box 1437
Oconee	11/27/59	262,176	104,511	40%		Gainesville 30501
Total		1,903,091	845,790	44%	2.3%	
ILLINOIS						
Shawnee	9/6/39	839,728	254,157	30%	.7%	317 East Poplar
						Harrisburg, Illinois 62946
INDIANA						
Hoosier	10/1/51	645,086	177,603	28%	.8%	1615 J Street
						Bedford, Indiana 47421
KENTUCKY						
Daniel Boone	2/23/37	1,357,086	520,038	38%		27 Carol Road
Jefferson	4/21/36	54,614	961	2%		Winchester, Kentucky 40391
(See also WEST VIRGINIA)						
Red Bird Purchase Unit		687,061	126,063	18%		
Total		2,098,761	647,062	31%	2.5%	
LOUISIANA						
Kisatchie	6/10/30	1,017,328	595,589	59%	2.1%	2500 Shreveport Highway
						Pineville, Louisiana 71360
MAINE						
White Mountain	5/16/18	81,316	45,944	57%	.2%	719 Main Street, P.O. Box 638
(See also NEW HAMPSHIRE)						Laconia, New Hampshire 03246

265

STATE	Date Established	Gross Acreage Within Boundaries	Net Acreage	Net as % of Gross	Net NF as % of Total State Area	Administration
MICHIGAN						
Hiawatha	1/16/31	1,281,668	863,885	67%		524 Ludington Street Escanaba, Michigan 49829
Huron	2/11/09*	695,337	418,348	60%		421 S. Mitchell Street
Manistee	10/25/38	1,331,585	497,099	37%		Cadillac, Michigan 49601
Ottawa	1/27/31	1,559,891	916,566	59%		Ironwood, Michigan 49938
Alpena S.F. PU		240	240			
Fife Lake S.F. PU		493	493			
Total		4,869,214	2,696,631	55%	7.4%	
MINNESOTA						
Chippewa	5/23/08*	1,599,631	655,623	41%		Cass Lake, Minnesota 56633
Superior	2/13/09*	4,011,731	2,153,352	54%		Duluth, Minnesota 55801
Total		5,611,362	2,808,975	50%	5.5%	
MISSISSIPPI						
Bienville	6/15/36	382,821	177,077	46%		National Forests in Mississippi
Delta	1/12/61	118,200	59,159	50%		Milner Building, Box 1291
De Soto	6/17/36	796,649	500,156	63%		Jackson, Mississippi 39205
Holly Springs	6/15/36	519,952	145,141	28%		
Homochitto	7/20/36	373,497	189,039	51%		
Tombigbee	11/27/59	119,155	65,412	55%		
Total		2,310,274	1,135,984	49%	3.8%	
MISSOURI						
Cedar Creek Purchase Unit		76,904	0	0		RDI Building, Suite 500
Mark Twain	9/11/39	2,991,650	1,439,034	49%		3003 E. Trafficway
Total		3,068,554	1,439,034	47%	3.3%	Springfield, Missouri 65802
NEW HAMPSHIRE						
White Mountain (See also MAINE)	5/16/18	805,144	683,637	85%	11.8%	719 Main Street, P.O. Box 638 Laconia, New Hampshire 03246
NORTH CAROLINA						
Cherokee (See also TENNESSEE)	7/14/20	327	327	100%		National Forests in North Carolina Plateau Building
Croatan	7/29/36	308,226	156,589	51%		50 S. French Broad
Nantahala	1/29/20	1,366,027	457,772	34%		P.O. Box 2750
Pisgah	10/17/16	1,076,511	483,154	45%		Asheville, North Carolina 28802
Uwharrie	1/12/61	220,202	45,760	21%		
Yadkin Purchase Unit		194,496	0			
Total		3,165,789	1,143,602	36%	3.6%	
OHIO						
Wayne	10/1/51	833,228	163,561	20%	.6%	1615 J Street Bedford, Indiana 47421
OKLAHOMA						
Ouachita (See also ARKANSAS)	12/18/07	412,912	244,489	59%	.6%	Box 1270, Federal Building Hot Springs National Park Arkansas 71901
PENNSYLVANIA						
Allegheny	9/24/23	742,693	506,102	67%	1.8%	Spiriden Building Warren, Pennsylvania 16365

STATE	Date Established	Gross Acreage Within Boundaries	Net Acreage	Net as % of Gross	Net NF as % of Total State Area	Administration
UTH CAROLINA						
ncis Marion	7/10/36	414,700	249,406	60%		National Forests in South Carolina
nter	7/13/36	965,762	357,599	37%		1813 Main Street, Room 350
Total		1,380,462	607,005	44%	3.1%	Columbia, South Carolina 29201
NNESSEE						
erokee	7/14/20	1,211,907	618,894	51%	2.3%	Federal Building, Box 400
also NORTH CAROLINA)						Cleveland, Tennessee 37311
KAS						
gelina	10/13/36	391,300	155,293	39%		National Forests in Texas
vy Crockett	10/13/36	394,200	161,478	41%		Federal Building, Box 969
ine	10/13/36	440,167	187,191	43%		Lufkin, Texas 75901
n Houston	10/13/36	491,800	158,648	32%		
Total		1,717,467	662,610	39%	.4%	
RMONT						
en Mountain	4/25/32	629,518	254,025	40%	4.3%	Federal Building, 151 West Street Rutland, Vermont 05701
RGINIA						
orge Washington	5/16/18	1,638,773	940,352	57%		210 Federal Building
also WEST VIRGINIA)						Harrisonburg, Virginia 22801
erson	4/21/36	1,587,735	656,530	41%		Carlton P.O. Box 4009
also WEST VIRGINIA nd KENTUCKY)						Roanoke, Virginia 24015
Total		3,226,508	1,596,882	49%	6.3%	
ST VIRGINIA						
orge Washington	5/16/18	157,152	100,006	64%		Department of Agriculture Building
also VIRGINIA)						Sycamore Street, Box 1231
erson	4/21/36	29,782	18,245	61%		Elkins, West Virginia 26241
also VIRGINIA nd KENTUCKY)						
nongahela	4/28/20	1,673,590	839,287	50%		
Total		1,860,524	957,538	51%	6.2%	
SCONSIN						
equamegon	11/13/33	1,049,347	839,565	80%		Federal Building Park Falls, Wisconsin 54552
olet	3/2/33	973,405	652,001	67%		Federal Building
Total		2,022,752	1,491,566	74%	4.3%	Rhinelander, Wisconsin 54501
tal Eastern National Forests		46,441,916	23,758,013	51%		

igures include established national forests and associated purchase units. The few eastern Land Utilization projects and national rasslands, as well as research and experimental areas, while technically part of the National Forest System, have been omitted hen not a part of a national forest.

hese National Forests were originally created from a nucleus of public domain land prior to the passage of the Weeks Act.

OURCES: *Establishment and Modification of National Forest Boundaries, A Chronological Record 1891-1973*, U. S. Department of griculture, U. S. Forest Service, Division of Engineering, October, 1973; *National Forest System, Areas as of June 30, 1975*, epartment of Agriculture, U. S. Forest Service.

PARTICIPANTS IN REGIONAL WORKSHOPS

Airlie House, Warrenton, Virginia, June 11-12, 1975
(Convened by The Conservation Foundation)

Stewart M. Brandborg, The Wilderness Society, Washington, D.C.
John Capell, LENOWISCO Planning District, Duffield, Virginia.
George C. Cheek, American Forest Institute, Washington, D.C.
Louis S. Clapper, National Wildlife Federation, Washington, D.C.
Arthur Davis, Western Pennsylvania Conservancy, Pittsburgh, Pennsylvania.
George Davis, Adirondack Park Agency, Ray Brook, New York.
Robert T. Dennis, Zero Population Growth, Washington, D.C.
Russell E. Dickenson, National Park Service, Washington, D.C.
William Duerr, Virginia Polytechnic Institute and State University, Blacksburg, Virginia.
Barry R. Flamm, Council on Environmental Quality, Washington, D.C.
Charles H.W. Foster, School of Forestry and Environmental Studies Yale University, New Haven, Connecticut.
The Reverend William B. Gable, Konnarock, Virginia.
James E. Giltmier, U.S. Senate Committee on Agriculture and Forestry, Washington, D.C.
Perry R. Hagenstein, New England Natural Resources Center, Boston, Massachusetts.
John F. Hall, National Forest Products Association, Washington, D.C.
Alexander Hamilton, Virginia Farm Bureau Federation, Richmond, Virginia.
Charles R. Hartgraves, U.S. Forest Service, Washington, D.C.
Samuel P. Hays, Sierra Club, Pennsylvania Chapter, Pittsburgh, Pennsylvania.
Raymond M. Housley, Jr., U.S. Forest Service, Washington, D.C.

William Hyde, Resources for the Future, Washington, D.C.
Alice Klavans, League of Women Voters of the United States, Washington,D.C
Dennis C. LeMaster, Society of American Foresters, Washington, D.C.
John R. McGuire, U.S. Forest Service, Washington, D.C.
Hugh Montgomery, Appalachian Regional Commission, Washington, D.C.
Gerald R. Mylroie, American Institute of Planners, Washington, D.C.
James Nelson, Pennsylvania Bureau of Forestry, Harrisburg, Pennsylvania.
Patrick F. Noonan, The Nature Conservancy, Washington, D.C.
Richard D. Pardo, American Forestry Association, Washington, D.C.
William M. Partington, Florida Environmental Information Center, Winter Park, Florida.
John A. Sandor, U.S. Forest Service, Milwaukee, Wisconsin.
Kenneth Scholz, U.S. Forest Service, Washington, D.C.
Maitland S. Sharpe, Izaak Walton League of America, Washington, D.C.
Arthur V. Smythe, Weyerhauser Company, Washington, D.C.
David Stahl, Urban Land Institute, Washington, D.C.
Fred Swaney, Massanutten Development Company, Harrisonburg, Virginia.
Edward T. Walters, Supervisor, Bath County, Virginia.
Robert E. Wolf, Congressional Research Service, Library of Congress, Washington, D.C.
Arthur D. Woody, U.S. Forest Service, Birmingham, Alabama.
John Yolton, United Auto Workers, Detroit, Michigan.

Nemacolin Inn, Uniontown, Pennsylvania, October 23-24, 1975 (Convened by the Western Pennsylvania Conservancy)

J. Edward Adams, Venango County Planning Commission, Franklin, Pennsylvania.
Donald R. Andrews, West Virginia Forest Management Practices Commission, St. Albens, West Virginia.
Raymond Benton, U.S. Forest Service, Allegheny National Forest, Warren, Pennsylvania.
Michael Brewer, The Academy for Contemporary Problems, Washington, D.C.
Suzanne Broughton, League of Women Voters of North Hills, Pittsburgh, Pennsylvania.
E. James Bryner, Pennsylvania Oil and Gas Association, Custer City, Pennsylvania.
Robert Burrell, West Virginia Highlands Conservancy, Morgantown, West Virginia.
John Butt, U.S. Forest Service, Allegheny National Forest, Warren, Pennsylvania.
John Carroll, Pennsylvania Department of Environmental Resources, Harrisburg, Pennsylvania.

Robert V. Clark, Pennsylvania Forestry Association, Mechanicsburg, Pennsylvania.
Thomas Clark, Forest Land Management Company, Morgantown, West Virginia.
Rodney Clay, West Virginia Department of Natural Resources, Charlestown, West Virginia.
J. A. Cochran, Consolidation Coal Company, Pittsburgh, Pennsylvania.
Marian Crossman, North Area Environmental Council, Pittsburgh, Pennsylvania.
William Curry, Laurel Highlands Conservation and Development Project, Johnstown, Pennsylvania.
Arthur A. Davis, Western Pennsylvania Conservancy, Pittsburgh, Pennsylvania.
Howard Dietz, Izaak Walton League, Richwood, West Virginia.
Lawrence Dietz, West Virginia Department of Natural Resources, Richwood, West Virginia.
Donald Girton, U.S. Forest Service, Wayne-Hoosier National Forest, Bedford, Indiana.
Walter C. Gumbel, Monongahela Power Company, Fairmont, West Virginia.
Clifford Ham, American Youth Hostels, Pittsburgh, Pennsylvania.
Samuel Hays, Sierra Club, Pittsburgh, Pennsylvania.
Keith Horn, Kojancic and Horn Consulting Engineers, Kane, Pennsylvania.
Roger Johnson, U.S. Forest Service, Monongahela National Forest, Elkins, West Virginia.
Roger Latham, Pittsburgh *Press*, Pittsburgh, Pennsylvania.
Richard Martyr, National Audubon Society, Harrisburg, Pennsylvania.
Helen McGinnis, Boalsburg, Pennsylvania.
Ralph Mumme, U.S. Forest Service, Monongahela National Forest, Elkins, West Virginia.
James Nelson, Pennsylvania Department of Environmental Resources, Harrisburg, Pennsylvania.
M. Graham Netting, Western Pennsylvania Conservancy, Pittsburgh, Pennsylvania.
John C. Oliver, III, Western Pennsylvania Conservancy, Pittsburgh, Pennsylvania.
David K. Rice, County Commissioner, Warren, Pennsylvania.
John P. Robin, Allegheny Conference on Community Development, Pittsburgh, Pennsylvania.
Sayre Rodman, Western Pennsylvania Conservancy, Oakmont, Pennsylvania.
Bruce Rubidge, Laurel Foundation, Pittsburgh, Pennsylvania.
John Sandor, U.S. Forest Service, Eastern Region, Milwaukee, Wisconsin.
Kenneth Scholz, U.S. Forest Service, Washington, D.C.

Eugene Schreve, WESTVACO, Covington, West Virginia.
Arthur W. Schuette, Koppers Company, Pittsburgh, Pennsylvania.
James Speice, The Hammermill Paper Company, Erie, Pennsylvania.
Anthony P. Suppa, Western Pennsylvania Conservancy, Pittsburgh, Pennsylvania.
Joshua C. Whetzel, Jr., Western Pennsylvania Conservancy, Pittsburgh, Pennsylvania.
Paul G. Wiegman, Western Pennsylvania Conservancy, Pittsburgh, Pennsylvania.
Joseph Womble, Hammermill Paper Company, Erie, Pennsylvania.
H. G. Woodrum, West Virginia Department of Natural Resources, Charleston, West Virginia.

Atlanta, Georgia, October 25-26, 1975
(Convened by the Florida Environmental Information Center)

James A. Altman, American Pulpwood Association, Jackson, Mississippi.
J. Earl Bailey, Sierra Club, Tuscaloosa, Alabama.
Sally Battle, League of Women Voters, Columbia, South Carolina.
John M. Bethea, Florida Division of Forestry, Tallahassee, Florida.
Linda Billingsley, The Georgia Conservancy, Atlanta, Georgia.
Alastair Black, Dougherty Associates, Atlanta, Georgia.
F. Joseph Butler, Georgia Kraft Company, Rome, Georgia.
Andy Caldwell, Central Florida Dog Hunters Association, Umatilla, Florida.
Carolyn Carr, Auburn, Alabama.
John Coates, Jr., Bureau of Outdoor Recreation, Washington, D.C.
Charles B. Davey, North Carolina State University, Raleigh, North Carolina.
Robert Eikum, DeLand, Florida.
Henry B. Fishburne, South Carolina Society of Consulting Foresters, Charleston, South Carolina.
C. Allan Friedrich, The Georgia Conservancy, Cumming, Georgia.
Roy C. Gandy, Southern Region, U.S. Forest Service, Atlanta, Georgia.
Trudy Huger, Garden Club of America, Atlanta, Georgia.
Albert F. Ike, Institute of Community and Area Development, University of Georgia, Athens, Georgia.
William M. Liebenow, Virginia Polytechnic Institute and State University, Blacksburg, Virginia.
Eleanor McCann, City of Auburn Planning Department, Auburn, Alabama.
Sid McKnight, Stone Mountain, Georgia.
Robert E. Marvin, Walterboro, South Carolina.
Robert A. Mason, Georgia Forestry Commission, Atlanta, Georgia.
Annie Laurie Matthews, North Carolina Land-Use Congress, Durham, North Carolina.

Jack L. May, School of Forest Resources, University of Georgia, Athens, Georgia.
Ted H. Meredith, Southern Forest Products Association, New Orleans, Louisiana.
Peter R. Mount, Southern Appalachian Multiple-Use Council, Leicester, North Carolina.
Betty Murray, The Georgia Conservancy, Atlanta, Georgia.
John V. Orr, U.S. Forest Service, Columbia, South Carolina.
William M. Partington, Executive Director, Florida Environmental Information Center, Winter Park, Florida.
B.J. Pavlovich, American Plywood Association, Atlanta, Georgia.
Don Percival, U.S. Forest Service, Tallahassee, Florida.
Cecil R. Phillips, The Georgia Conservancy, Atlanta, Georgia.
Paul Pritchard, Appalachian Trail Conference, Harpers Ferry, West Virginia.
Robin D. Shaddox, U.S. Forest Service, Tallahassee, Florida.
Robert W. Simons, Alachua Audubon Society, Gainesville, Florida.
J. Owens Smith, Institute of Natural Resources, University of Georgia, Athens, Georgia.
Fred W. Stanberry, Florida Game and Fresh Water Fish Commission, Tallahassee, Florida.
Klaus Steinbeck, School of Forest Resources, University of Georgia, Athens, Georgia.
H. Daniel Stillwell, Appalachian State University, Boone, North Carolina.
David A. Stock, Stetson University, DeLand, Florida.
E. Amos Sumner, Florida State Senate Committee on Agriculture, Tallahassee, Florida.
Jim Thompson, Community Action Program Council, United Auto Workers, Smyrna, Georgia.
Victoria Tschinkel, Florida Department of Environmental Regulation, Tallahassee, Florida.
Walter R. Tschinkel, Florida State University, Tallahassee, Florida.
Susan Whisnant, School of Forest Resources, University of Georgia, Athens, Georgia.

Waterville Valley, New Hampshire, October 30-31, 1975. (Convened by New England Natural Resources Center)

Sherman Adams, Lincoln, New Hampshire.
Maurice D. Arnold, Bureau of Outdoor Recreation, Philadelphia, Pennsylvania.
Norman A. Arseneault, U.S. Forest Service, Green Mountain National Forest, Manchester Center, Vermont.
Richard Barringer, State Bureau of Public Lands, Augusta, Maine.
William Bentley, Harvard Forest, Petersham, Maine.

Robert Binnewies, Maine Coast Heritage Trust, Bar Harbor, Maine.
Paul O. Bofinger, Society for the Protection of New Hampshire Forests, Concord, New Hampshire.
John Bork, Brown Company, Berlin, New Hampshire.
Arthur F. Brown, Jr., Carrol County Trust Company, Conway, New Hampshire.
Richard H. Burt, Allen-Rogers Corporation, Laconia, New Hampshire.
Robert Butler, U.S. Forest Service, Green Mountain National Forest, Manchester Center, Vermont.
John C. Calhoun, Jr., Keene, New Hampshire.
James Carlaw, International Paper Company, South Glen Falls, New York.
Richard Clark, Selectman, Ripton, Vermont.
Hope Cobb, Appalachian Mountain Club, Princeton, New Jersey.
Thomas Corcoran, Waterville Company, Waterville, New Hampshire.
William Damon, Norway, Maine.
Hugh C. Davis, Institute for Man and His Environment, University of Massachusetts, Amherst, Massachusetts.
Alexandra Dawson, Conservation Law Foundation, Boston, Massachusetts.
Thomas Deans, Appalachian Mountain Club, Boston, Massachusetts.
Paul T. Doherty, Bureau of Off-Highway Recreational Vehicles, Concord, New Hampshire.
Ernest Gould, Jr., Harvard Forest, Petersham, Maine.
Perry R. Hagenstein, New England Natural Resources Center, Boston, Massachusetts.
George T. Hamilton, State Department of Parks and Recreation, Concord, New Hampshire.
John Hibbard, Connecticut Forest and Park Association, East Hartford, Connecticut.
Robert Hicks, Bristol, Vermont.
Fred E. Holt, State Bureau of Forestry, Augusta, Maine.
Lloyd Irland, School of Forestry and Environmental Studies, Yale University, New Haven, Connecticut.
Andrew Johnson, The A. Johnson Company, Bristol, Vermont.
Fred B. Knight, School of Forest Resources, University of Maine, Orono, Maine.
Paul Leavitt, Town Manager, Waterville Valley, New Hampshire.
Stephen J. Maddock, Appalachian Mountain Club, Boston, Massachusetts.
Howard Mason, Peck Lumber Company, Westfield, Massachusetts.
Kent Mays, U.S. Forest Service, Green Mountain National Forest, Rutland, Vermont.
Malcolm McLane, Orr and Reno, Concord, New Hampshire.
C. Edwin Meadows, Seven Islands Land Company, Bangor, Maine.
George Nagle, Adirondack Park Agency, Ray Brook, New York.

Paul Natale, U.S. Forest Service, White Mountain National Forest, Plymouth, New Hampshire.

Theodore Natti, State Department of Resources and Economic Development, Concord, New Hampshire.

Priscilla Newbury, New England River Basins Commission, Boston, Massachusetts.

David Newhouse, Schenectady, New York.

Andrew W. Poulsen, Littleton, New Hampshire.

Carl H. Reidel, University of Vermont, Burlington, Vermont.

Sheafe Satterwhite, Williams College, Williamstown, Massachusetts.

Judith T. Smith, American Association of University Women, North Conway, New Hampshire.

Seward Weber, Vermont Natural Resources Council, Montpelier, Vermont.

Paul D. Weingart, U.S. Forest Service, White Mountain National Forest, Laconia, New Hampshire.

Brendan Whittaker, Vermont Agency of Environmental Conservation, Montpelier, Vermont.

James E. Wilkinson, Jr., Vermont Agency of Environmental Conservation, Montpelier, Vermont.

Carl N. Wilson, U.S. Forest Service, Milwaukee, Wisconsin.

index

Quotations and captions are indicated by *italics*.

A

Acquisition. *See* Land acquisition
Adirondack Park, N.Y., 83, 179-180
Agricultural Conservation Program, 111
Agricultural land, protection, 165-166, 250
Alabama National Forest, Ala., 15
Allegheny National Forest, Pa. 15, *16*, 44, 236
   Current uses, 255-257
   Forest plan, 123, 128
   Future vision, 257-262
   Land acquisition, 205
   Map, 255
   Mineral rights, 58, *60*, 200, 202
   Oil drilling, 4, 57, *59*, 61, *174*, 176, *176*, 177, *209*, 210, *210-211*, 215
   Revenue sharing, 227, 228
   Timber, 3, *23*, 256
   Visitors, 149, 256
Allegheny River Boat Trail, Pa., 261
Amenities. *See* natural amenities
American Forestry Association, 12
Angeles National Forest, Calif., 5
Apalachicola National Forest, Fla., 16
   Timber, 35
Appalachian Mountain Club, 238-239, *240*, 243-244
Appalachian Mountains
   Land ownership patterns, 8
   National forests, 1, 13, 15, 17
   Planning area, 132
   Strip mining, 213
   Timber, 39, 93
   See also individual forests
Appalachian Trail, *43*, 82, 85
   Land acquisition, 219
Arkansas
   National forests, 8
   Stand conversion, 35
Arrowhead Regional Development Commission: Minnesota, 163, 171

B

Beaverhead National Forest, Mont., 168
Big Island Lake, Mich., 100
Boundary Waters Canoe Area, Minn., 46, 47, 61-62, 163, 227
Bristol Cliffs Wilderness, Vt., *166-168*
Bureau of Land Management, land acquisition funds, 221
Bureau of Mines, 211
Byrd, Robert, 223

C

Camping, 42, 44, 72
Campgrounds, 41, 94, 158
   Privately owned, 94, 112, 251-252, 259
   User fees, 44, 108
Cannon, Joseph, 14
Cape Cod National Seashore, Mass., 178-179, 181, 185
   Advisory commission, 190-193, 194
Cappaert v. United States, 186
Chattahoochee National Forest, Ga., 44, *191*
   Land acquisition funds, 219
Chattooga River, Ga., 63
Cheoah Bald, N.C., 100-101
Chequamegon National Forest, Wisc., 16
   Land acquisition funds, 218
   Visitor days, 42
Cherokee National Forest, N.C.-Tenn., 15
   Land acquisition funds, 219
Chippewa National Forest, Minn., shared revenues, 228
Civilian Conservation Corps, 3, 16, 105, *105*, *114*
Clark National Forest, Mo., 16
   Revenue sharing, 227
Clarke-McNary Act (1924), 15, 120
Clawson, Marion, 123, 139
Clearcutting, 28, *29*, 30, 41, 93, *123*, 125, 248
Cleveland National Forest, Calif., 5
Coal mining, 19, 58, 60, 62, *175*, 176, 202, 211, 212-214
Coastal Plain, 44
   Forests, 1, 2, 92
Coastal-zone management plans, 165

277

Cohutta Wilderness, Ga.-Tenn., 46
Colorado, national forests, 5, 8
Commercial forest, 26, 30, 90. *See also* Timber production
Condemnation authority, 181, 185, 186, 206
Conservation, 17, 80, 122. *See also* Sustained-yield management
Cooperative action, 147, 157-194, 199, 239-240, 246-247, 260-262
  Agreements, 168-171
  Coordinating agencies, 179-180
  Federal advisory committees, 189-194
  Funding, 177-178
  Need for, *160-161*
  Obstacles to, 159-162, 225-26
  Public involvement, 187-188
Copeland Report, 204
Craig County, Va., national forest receipts, 230
Cranberry Backcountry Wilderness Area, Va., mineral rights, 58, 60
Critical areas legislation, 164-165, 184

D

Daniel Boone National Forest, Ky., 15
  Land acquisition, 206, 218
  Mineral rights, 58
  Mining, 176
Davis, Deane C., 164
Demand, 4-5, 71, 79
  Conflicts, 4-5, 18, 19-21, 143
  Future, 35-41, 49-50, 64, 80-81, 149
  Increasing, 18, 51
Demonstration forests, 18, 111, 172, 208
DeSoto National Forest, Miss., 92
  Oil production, 57
Development, 64-65, 85, 157-158
  Controls, 96, 160, 165, 185
  Cooperative planning, 169, 240
  Environmental impact, 19, 51-52, 149, 202
  Patterns, 85, 201
  Recreational, 44-45, 51-52, 61
Development rights, 166, 216-217
Devil's Hole National Monument, Nev., 186
Dittmer, Kenneth R., 182

E

Easements
  Scenic, 182, 183, 184, 216-217, *217*
  Trail, 112
Eastern Wilderness Areas Act (1974), 46, 100
Ecosystems, 56, 82-83
  Management, 97, 246-248, 254
Eglin Air Force Base, Fla., 18, 63
Environmental impact
  Development, 19, 51-52, 149, *209*
  Mining, 58, 61-62
Environmental quality, maximizing, 20, 87-91, 125, 127, 129, 132, 235
Erosion, 54-55
Evans, Frank, 229
Even-aged timber management, 28
Everglades National Park, Fla., 83

F

Federal Advisory Committee Act (1972), 189-190, 194
Federal advisory committees, 189-194
  Weaknesses, 193
Federal Government
  Cooperative action, 180-181, 183-184, 187
  Land-use constraints, 185-187
  Mining legislation, 76, 176, 212-225
  Pollution control laws, 160, 176
Fire Island National Seashore, N.Y., 181, 185
Fish and Wildlife Service, land acquisition funds, 221
Flooding, 14, 54
  Control, 256
Florida, land-use laws, 163-164
Florida Committee on Rare and Endangered Plants and Animals, 252
Ford, Gerald R., 139, 213
Forest and Rangeland Renewable Resource Planning Act of 1974 (RPA), 40-41, 44, 119, 124-125, 126, 136-140, 160
  Assessment, 138-140, 212, 240
  Management, 121, 124-125, 129, 136-140
Forest management. *See* Land-management planning; Multiple-use management; Timber production
Forest products. *See* Wood products
Forest Service, *123*, *137*, 141-153, *143*
  Budget appropriations, 107, 119, competition for, 20
  Cooperative action, 161,168-174
  Educational role, 94-95, 110-111
  Founding and history, 11-13, 141
  National Forest System branch, 144-146, 171-172, 178
  Objectives, 12, 25, 130
  Organization, 143-148
  Personnel, 148-153
  Research branch, 144
  Roles and responsibilities, 141-143, 173
  State and Private Forestry branch, 12, 110, 144, 146-148, 171-172, 177-178
Forest users, *26*
  Consumers, 65, 70-71, 127-128; profile, 70, 72
  Producers, 65-70
Forestry Incentives Program, 111
Francis Marion National Forest, S.C., 92
Frome, Michael, 143
Future
  Demands, 35-41, 49-50, 64, 80-81, 149
  Potential uses, 64-65, 89, 235-263
  Preservation for, 88-89, 104

G

Gee Creek Wilderness, Tenn., 46
General Land Office, 12
Geological Survey, 12, 55, 103
George Washington National Forest, Va., 15, 64, 123, 162, 168, 221
  Development, 157, 202
  Land acquisition funds, 219
  Mineral extraction, 55, 62
  Private land, 4
  Wildlife, 52
Giltmier, Jim, 188
Graves, Henry S., 204

Grazing, 56-57, 102-103, *139*
  Fees, 27, 56
Great Depression, 16, 18
Great Gulf Wilderness Area, N.H., 46
Great Smoky Mountain National Park, N.C.-Tenn., 83
Green Line Parks, 183-184
Green Mountain National Forest, Vt., 5, 16, 164
  Budget, 145
  Cooperative action, 169-171
  Land acquisition, 223, *225*
  Staff, 145
  Wilderness areas, *167-168*
  Wildlife, *98*
Green Mountains, Vt., 85
Guerrero, Walter A., 224

H

Hadley, Lawrence, 194
Hardwoods, 26-27, 81
  Consumption, 38-39
  Forage producers, 57
  Forestry practices, 32, 41, 93
  Production incentives, 39, 40
Hells Canyon National Recreation Area, Idaho, 181
Herfindahl, Orris, 122
Hiawatha National Forest, Mich., 57, 100
Holly Springs National Forest, Miss., 201, 202
Homochitto National Forest, Miss., 92
  Oil production, 57
  Revenue sharing, 227
Hoosier National Forest, Ind., 16, 41
  Land exchange, 223-224
  Ownership patterns, 201
Housley, Raymond M., 152
Housing, timber demand, 35, 37, 40
Humphrey, Hubert, 136
Hunting and fishing, 44, *53*, 53, 98-99, *99-100*, 254, 257-258
  Income generation, 68-69
Hunting rights, leased, *99*, 112, 217, 254
Huron National Forest, Mich., 52

I

Idaho, national forests, 8
Illinois, national forests, 8
Incentives, 106-114
  Local governments, 112-114
  Private landowners, 106, 110-112, 242, 250
  Public land managers, 106
Indiana, national forests, 8
Inland Steel Corporation, 61
International Nickel Company, 61
Interpretation, 94-95, 252
Isle Royale National Park, Mich., 83
Itasca County, Minn., national forest receipts, 228

J

Jefferson National Forest. *See* Thomas Jefferson National Forest
Job Corps, *21*, *105*

Jobs programs, 104, 105-106,*106*
Johnson, Bennett, 184

K

Kansas, national forests, 8
Kaufman, Herbert, 141, 146, 152
Kennedy, Edward, 184
Kirtland's Warbler Management Area, Mich., 52
Kisatchie National Forest, La., 91
  Land exchange, 224
  Military use, 4, 63
  Mineral deposits, 62
  Revenue sharing, 227
  Timber, 2, 3
  Visitor days, 42

L

Lake Conroe Reservoir, Tex., 63
Lake States, 1
  Grazing land, 57
  Hardwoods, 93
  Ownership patterns, 8
  Timber production, 39
Land
  Acquisition, 15-16, 55, 120-121, 199-201, 203-208, 222, 223, 250-251; advantages, 69; condemnation, 185, 186; criteria for future, 207-208; funds, 13, 15, 141, 217-222; justification, *14*; less-than-fee interests, 181-182, 183-184, 216-217; local tax base and, *69*, 205, 223; mineral rights, 200, 201-202, 208-216
  Exchange, 223-225; interstate, 224-225; trading stock, 208
  Ownership patterns, 2-3, 5, 8, 19, 84, 127, 199, 201-203
  Prices, 218, 219, *219*, amenity value, 223
  Rehabilitation, 3-4, 87-88, 104-105, 200-201, 206, 209, 216, 235
  Stewardship, 90, 111, 130-131
  See also Private land
Land and Water Conservation Fund Act, 55, 201
  Appropriations, 217-222
Land-management planning, 131-153, 157-194
  Adjacent private lands, 134, 135-136
  Integration of efforts, 133, 135-136, 150-151, 168-171, 173-174, 184
  Forest plans, 132, 133-136
  Public involvement, 171, 187-193
Land-use controls, 136, 159-162, 164-166, 199
  Cape Cod formula, 180-181
  Enforcement, 185-187
Linville Gorge Wilderness Area, N.C., 46, 47
Local economy, 48
  Forest impact on, 2, 20, 67-70, 112-113, *113*, 135-136, 158-159, 230, 242, 255
  Tax base and federal land acquisition, *69*, 205, *205*, 223
  See also Revenue sharing
Local government
  Cooperative action, 169-171, 178, 184, 239-240, 260
  Federal land acquisition, attitude toward, 205, 206
  Forest benefits, 230
  Incentives, 112-114
  Land-use controls, 136, 199; resistance to, 161-162

Los Padres National Forest, Calif., 5
Low management intensity areas, *101*, 101, 158, 259

M

McArdle, Richard E., 121
McGuire, John, 3, 45, 147, 148, 230
McKim, C. R., 14
McKinley, William, 13
Madden, Carl H., 160
Maine, national forests, 8
Mark Twain National Forest, Mo., 16, 83
    Mining, 57, 61
    Staff and budget, 146
Maroon Bells-Snowmass Wilderness Area, Colo., 82
Massachusetts, forest management, *84*, 147
Massanutten Mountain, Va., *114*, 123, 157, 202
Michigan, national forests, 8
Military use, 4, 18, 63
Mineral rights, 2, 16-17, 58, *60*, 62, 209, 256, 262
    Acquisition, 200, 201-202, 208-216, *209*
    Valid existing rights, 214-215
Mining, 4, 57-62
    Environmental effects, 2, 19, 55, 56, 58, 61-62
    Federal mineral leases, 60, 103, *209*
    Regulation of, 58, 60, 103-104, 176, 210, 211-215, 262
Minnesota, national forests, 8
Missouri, national forests, 16
Mitchell, Mount, N.C., 85
Monongahela National Forest, W. Va., 3, 14-15, 79, 162
    Clearcutting, 30, 55, 120, 158, 193
    Coal mining, 210-211, 125
    Land acquisition, 206, 218, 223
    Mineral rights, 58, 202
Monongahela River, 1907 flood, 14-15
Mount Baker National Forest, Wash., 5
Mount Rogers National Recreation Area, Va., 56, *139*, 163, 206, 219
Multiple-use management, 2, 17-18, *18*, 25, *26*, 31, 121-124, 131, 248
    Maximizing unique public benefits, 79, 85-86, 88, 207, 235, 242
Multiple-Use Sustained Yield Act (1960), 17, 25, 119, 121-124, 128, 129

N

Nantahala National Forest, N.C., 15, 169, *188*
    Development in, 157
    Land acquisition funds, 219
    Wilderness area, 100-101
National Environmental Policy Act, 119, 188
National Forest Management Act (1976), 30, 119, 120, 124-125, 129, 130, 136
National forests
    Budget, 145
    Eastern: acreage, 1, 5, 8, 83-84; compared with west, 5-9, 80-83; history, 3, 9-10, 13-16, 87; special characteristics, 83-85, 126-127; statistics, 264-267
    Legislation, 10-11, 13, 15, 119-131
    Roles and management, 8-9, 17-21, 120, 129
    Western, 5, 8; timber, 81-82
    See also Forest Service; *and* individual areas of management
National Park Service, 186
    Advisory commissions, 190-192
    Land acquisition, 220-221
    Mission, 130
National parks, 17
    Land-use regulation, 180-181
    Size, 83
National Wilderness Preservation System, 256. *See also* Wilderness areas
Natural amenities, 85, 89, 94, 123, 127-129
Natural Bridge National Forest, Va., 15
Nature Conservancy, 206, 223
New England, national forests, *11*, 102
New England River Basins Commission, *160-161, 164*
New Hampshire, national forests, 8
New Mexico, national forests, 8
Nevada, national forests, 5
New York, land-use legislation, 166. *See also* Adirondack Park
Nicolet National Forest, Wisc., 16, 202
    Development, 157
Nonrenewable resources, 35
    Exploitation, 103-104
North Carolina, 40
    Timber industry, 3
North Country Trail, Pa., 261

O

Ocala National Forest, Fla., 236
    Clearcutting, *29*
    Current uses, 244-246
    Development in, 135, 157, *193*, 202, 224, *246*
    Future vision, 246-254
    Land exchange, 224
    Map, 245
    Military use, 63
    Private inholdings, 158
Oconee National Forest, Ga., 2, 208
Off-road vehicles, 47, *47-49*
Office of Management and Budget, 151
Ohio
    National forests, 8
    Strip mining laws, 176
Oil drilling, 4, 19, 57, *59*, 61-62, *174*, 176, *176, 177*, *209*, 210, *210-211*, 215, 256
Older Americans Act employment program, 105
Open space, 20, 84-85, 221
Organic Administration Act (1897), 11, 30, 119-120, 128, 129
Oregon, national forests, 8
Oregon Dunes National Recreation Area, Oreg., 181
Osceola National Forest, Fla., 16, 79, 91
    Mining, 60
Ottawa National Forest, Mich., advisory committee, 190
Ouachita Mountains, 85
Ouachita National Forest, Ark., 4, 83
    Coal deposits, 62

Ozark Mountains, 85
   Timber stand conversion, 40
Ozark National Forest, Ark., 83
   Land acquisition, 218

P

Pennsylvania
   Cooperative action, 260-262
   Environmental laws, 176
   Forest lands, 259
   Hunting, 257-258
Pennzoil Corporation, 176, *177*
Pinchot, Gifford, 10, 11-12, 17, 18, 25, 128, 146
Pisgah National Forest, N.C., 15, *192*
   Land acquisition funds, 219
Pleasure driving, 44, 95
Pollution, water, 19, 51, 54-55, 56, 61
   Regulation, 102, 160, 176
Preservation, 63, 87
   Future options, 88-89, 104
President's Panel on Timber and Environment (1973), 40
Private land, 2, 4, 158-159, 204
   Assistance to, 12, 146-147
   Constraints on, 167-168, 185-187, 207
   Cooperative action, 157
   Development, 158, 202
   Incentives, 90, 99, 106, 110-112, 166, 236
   Interaction with national forest, 19, 149, 157-158
   Land management, 85, 134, 135-136
   Ownership patterns, 201
   Public benefits, 85, 207
   Recreational facilities, 45, 52, 94, 96, 112-113, 251-252, 259
   Recreational residences, 45, 52
   Timber production, 31, 39, 89-91, 92, 110, 166, 242, 249-250, 259-260
Public Land Law Review Commission, 107, 204, 226
Public participation, land-use planning, 187-189. *See also* Federal advisory committees
Pulpwood, 31, 37, 39, 91
   Clearcutting, 41
   Pulp mills, 67

R

Rangelands, 56-57, 102-103
Recreation, 1, 4, 18, 20, 41-52, 82, 94-96, 256
   Activities, 41, 42, 44, 50-51, 71
   Budget appropriations, 107
   Consumer profile, 70-72, 127-128
   Developed, 44-45, 51-52, 61
   Dispersed, 44, 94, 252
   Facilities, 41, 44-45, 51-52, 94, 244; private land, 45, 52, 94, 96, 112, 251-252, 259
   Future demands, 49-52
   Growth rate, 43
   Land acquisition, 218-220
   Use conflicts, 47, 51
   Use restrictions, 47, 93, 243-244
   User fees, 44, 70, 106, 107-110
   Visitor days, 41-43, 68, 107
Recreational residences, 45, 70, 164, *192*
Redbird Purchase Unit, Ky., 206, 218
Redwood National Park, Calif., 186-187

Regional planning, 132-133, 135-136, *160*, 163, 174
   Cooperative action, 170, 262
   Renewable resources, 139-140
Renewable resources, 80, 149
   Systems, 19, 138; *see also* Sustained-yield management
Research, 12, 144
Reservoirs, 63-64
Resource Planning Act. *See* Forest and Rangelands Renewable Resource Planning Act
Revenue Sharing, 11, 69, 112, 121, 159, 225-231, 262
   Inequities, 160, 226, 227-229, 230-231
   Reforms, 229-230
Rice, David K., 228
Rivers, 55
   Management, 164
   Regional planning, 163
Roads, 19, *100*, *177*, 243, 248
   Maintenance, 28, 30
   Recreational use, 44, 95
Rockefeller, Nelson, 179
Roosevelt, Theodore, 12
Rotation age, 28, 31
   Hardwoods, 93

S

Sabine National Forest, Tex., 44
St. Louis County, Minn., 162
Sam Rayburn Reservoir, Tex., 63
San Bernadino National Forest, Calif., 5
Santa Monica Mountains and Seashore Urban Recreation Area, Calif., 184
Sawtimber, 26, 31, 37, 91-92, 249
Sawtooth National Recreation Area, Idaho, 179, 181-183, 185, 216
   Land acquisition, 182-183
Selective cutting, 30, 41, 93
Shawnee National Forest, Ill.
   Mining, 57
   Ownership patterns, 201
Shenandoah National Park, Va. 221
Shining Rock Wilderness, N.C., 46
Sierra Club v. Department of Interior, 186
Siuslaw National Forest, Oreg., revenue sharing, 227
Snoqualmie National Forest, Wash., 5
Snowmobiles, 47, *47-49*
Society of American Foresters, 122, 157
Society for the Protection of New Hampshire Forests, 13, 240
Softwoods, 27, 41, 81-82
   Consumption, 38-39
   Stand conversion, 34-35
Southern Forest Resource Council, 39
Southern pine forests, 26-27, 39, 92
   Grazing, 56
   Stand conversion, 34-35
Spruce Knob-Seneca Rocks Recreation Area, W. Va., 206, 219, 223
State governments
   Cooperative action, 147, *174-178*, 184, 240, 260-262

281

Land-use planning, 159, 160, 163-166, 174
Mining regulation, 174, 174-177, 177, 213
Strip mining, 60, 62, 175
  Land restoration, 104-105, 200-201, 206, 209, 216
  Regulation, 176, 211-216
Superior National Forest, Mich., 57, 83, 163, 171, 227
  Advisory committee, 190
  Mining, 61
Supreme Court, 186
Sustained-yield management, 31, 80, 121, 122

T

Talladega National Forest, Ala., 132, 208
Teton National Forest, Wyo., 168
  Revenue sharing, 227
Thomas Jefferson National Forest, Ky.-Va.-W. Va., xxx, 15, 54, 139, 162, 221
  Grazing, 56
  Land acquisition funds, 219
  Mineral rights, 58, 62
Thornton, Philip, 147
Timber production, 2, 15, 18, 20, 26-41, 65-67, 79-80, 81, 88, 91-93, 137, 256, 260
  Costs, 28
  Cutting practices, 28, 29, 30, 32-33, 41, 93, 120, 125, 248
  Future demand, 35-41
  Intensive management, 39-40, 82, 91
  Local dependence on, 3, 66, 67
  Logging practices, 10, 23, 32-33, 34
  Old growth stocks, depletion, 36, 39, 81-82, 112
  Prices, 112
  Private lands, 31, 39, 89-91, 92, 110, 242, 249-250, 259-260
  Rotation age, 28, 31, 93
  Stand conversion, 34-35, 39-40, 54, 92
  Timber sales, 27-28, 31, 40; revenues, 27-28, 107, 109
  Wildlife management, 26, 34, 54, 97-98
Toledo Bend Reservoir, Tex., 63
Transfer Act (1905), 12
Trust areas, 184

U

Udall, Morris K., 213
Urban centers, forest needs, 1, 5, 19, 20, 64, 70
User fees, 27, 44, 56, 71, 106, 107-110, 112, 259
Uwharrie National Forest, N.C., 16
  Ownership patterns, 201

V

Vermont, land-use planning laws, 163, 169-171
Virginia
  Land-use legislation, 162, 164-165, 166
  National forests, 8
Voyageurs National Park, Minn., 163, 171

W

Warren County, Pa.
  National forest receipts, 227, 228
  Public lands, 205
Washington, national forests, 5
Washington, Mount, N.H., 85, 238

Water, quality, 56
Watershed
  Effect of development on, 19
  Management, 54-56, 102, 103, 164, 246-248; legislation, 120
  Protection, 15, 18
Wayne National Forest, Ohio, 2, 16, 42
  Land acquisition, 200, 208, 218, 220
  Map, Marietta unit, 198
  Mineral rights, 202
  Mining, 55, 58, 62, 104, 175, 176, 211
  Ownership patterns, 200, 201, 208
Weeks Act (1911), 15, 54, 55, 119, 120-121, 126, 203-204
  Appropriations, 200, 217-220
Weingart, Paul D., 190
West Virginia
  National forests, 8, 15
  Strip mining, 214
  Zoning laws, 162
West Virginia Forest Management Practices Commission, 123
Western Pennsylvania Conservancy, 205, 223, 261
Whiskeytown-Shasta Trinity National Recreation Area, Calif., 181
White Mountain National Forest, Me.-N.H., 2, 3, 15, 17, 42, 100, 123, 135, 234, 236
  Advisory committee, 190
  Current uses, 237-239
  Establishment, 14
  Future vision, 237-244
  Land acquisition, 206
  Map, 237
  Ownership patterns, 201
  Timber, 26, 32, 67
White Mountains, N.H., 3, 15, 85
Wild and Scenic Rivers, 248
Wild River, N.H., 100
Wilderness Act (1964), 45, 47, 126
Wilderness areas, 45-47, 50-51, 80, 100-101, 167-168, 256, 258-259
  Management, 46-47
  Purity issue, 45-46, 101
  Study areas, 46
Wildlife, 9, 54, 98
  Endangered species, 52, 88, 98
  Nonconsumptive uses, 97-98
  Preservation, 63, 86
  Management, 52-54, 96-99, 252, 254; game animals, 52-53; habitat maintenance, 96-97
  Timber production and, 26, 34, 54, 97-98
William B. Bankhead National Forest, Ala., 15
Wilson, James, 13, 17, 128
Wood products, 1-2, 35, 65
  Consumers, 71
  Demand and price, 35-37, 40
  Substitutes, 35-36, 67
World War II, 18, 63
Wyoming, national forests, 5, 8

Y

Youth Conservation Corps, 105

Z

Zoning, 157, 160, 160, 162-163, 171, 185

This Conservation Foundation report focuses on the 50 national forests east of the Rocky Mountains—the largest public land system in the most urbanized section of the United States. It examines their characteristics and uses, the demands made upon them, the threats they face, and the potential they offer to the entire nation.

"THE LANDS NOBODY WANTED is a study everyone interested in the future of America's natural heritage should read. It presents a dramatic conservation success story, but it is also a warning and a challenge to those interested in land-use policy, forestry, recreation, and the wise use of the nation's resources."

Charles H. W. Foster, *Dean*
School of Forestry and Environmental Studies
Yale University

"THE LANDS NOBODY WANTED will become a benchmark for future forest policy in the East. This comprehensive study should be required reading for all students of forest policy and for all who make public lands policy. From history and origin to the direction future management should take, this book completely and accurately dissects the eastern national forest system. Because of their location with respect to human populations and needs, these forests may easily become the most valuable public lands in America. Regardless of whether or not we agree with all the conclusions of the report, no one can dispute the fact that The Conservation Foundation has done a great service in focusing attention on this public land asset."

William E. Towell, *Executive Vice President*
The American Forestry Association

THE CONSERVATION FOUNDATION
1717 Massachusetts Avenue, N.W., Washington, D.C. 20036